Thought Experiments in Methodological and Historical Contexts

History of Science and Medicine Library

VOLUME 19

Medieval and Early Modern Science

VOLUME 15

Thought Experiments in Methodological and Historical Contexts

Edited by

Katerina Ierodiakonou
Sophie Roux

BRILL

LEIDEN • BOSTON
2011

Cover illustration: Camille Flammarion, *L'atmosphère: météorologie populaire* (1888). Anonymous woodcut. Colorization by Frédéric Schmitt.

This book is printed on acid-free paper.

Library of Congress Cataloging-in-Publication Data

Thought experiments in methodological and historical contexts / edited by Katerina Ierodiakonou, Sophie Roux.
p. cm. – (History of science and medicine library, ISSN 1872-0684 ; v. 19) (Medieval and early modern science ; v. 15)
"This book is the distant outcome of a workshop on thought experiments that took place in Athens in April 2007"–Acknowledgements.
Includes bibliographical references and index.
ISBN 978-90-04-20176-7 (hardcover : alk. paper)
1. Thought experiments–History–Congresses. I. Ierodiakonou, Katerina. II. Roux, Sophie. III. Title. IV. Series.

BD265.T46 2011
190–dc22

2011011640

ISSN 1872-0684
ISBN 978 90 04 20176 7

CONTENTS

PART THREE

HOW DO THOUGHT EXPERIMENTS WORK?

ACKNOWLEDGEMENTS

This book is the distant outcome of a workshop on thought experiments that took place in Athens in April 2007. This workshop, co-organised by Frans De Haas, Katerina Ierodiakonou and Sophie Roux, was part of the programme "From Natural Philosophy to Science" generously sponsored by the European Science Foundation. In addition to the scholars whose essays follow in this volume, participants at the workshop also included Dionyssis Anapolitanos, Hugues Chabot, Vasso Kindi, Gerhard Krieger, Remke Kruk, Inna Kupreeva, Cees Leijenhorst and Chris Martin. We would like to thank the graduate student Ioannis Papachristou for his assistance with the practicalities of the meeting.

The Athens workshop was followed by an editorial meeting in Grenoble in June 2008, where a selection of papers read in Athens in preliminary form, once thoroughly revised, was keenly discussed. John Zeimbekis kindly agreed to present a paper at this meeting without having been present in Athens. For her efficiency in helping to organise Grenoble's editorial meeting, we thank Loredana Truong, administrator of the group "Philosophie, Langage et Cognition" of the University of Grenoble.

For their constant support, we thank Hans Thijssen, Chairman of the programme "From Natural Philosophy to Science," and Cees Leijenhorst, Coordinator of the same programme. Finally, this book would not have been possible without Birgit Kolboske, who helped us in the last phase of a long editing process.

LIST OF CONTRIBUTORS

Pascal Engel, *University of Geneva, Switzerland*
Jean-Yves Goffi, *University of Grenoble, France*
Christophe Grellard, *University of Paris I, France*
Katerina Ierodiakonou, *University of Athens, Greece*
Simo Knuuttila, *University of Helsinki, Finland*
Taneli Kukkonen, *University of Jyväskylä, Finland*
Peter Lautner, *Pázmány Péter Catholic University, Hungary*
Carla Rita Palmerino, *Radboud University Nijmegen, The Netherlands*
Sophie Roux, *University of Grenoble, France*
Stelios Virvidakis, *University of Athens, Greece*
John Zeimbekis, *University of Grenoble, France*

INTRODUCTION: THE EMERGENCE OF THE NOTION OF THOUGHT EXPERIMENTS[1]

Sophie Roux

In Internet forums for model makers, one sometimes finds discussion threads entitled "Luftwaffe 1946." Participants in this scenario imagine that World War II did not end in 1945, but continued into 1946, and that the German jets, which in 1944 or 1945 were only projects or prototypes, were operational. Given this scenario, they then describe the aerial combats that could have taken place between these planes and similar ones developed by the Allies as if they had actually taken place (the most gifted among the forum members accompany their posts with very spectacular computer-generated images). A minimum of historical realism is required: one cannot imagine, for example, a mutiny of the Soviet generals, who would eliminate Stalin and demand a separate armistice. We note that a scenario of this type is all the more credible when the person who imagines it has good knowledge of aerial warfare and the aeronautics industry. The last notable element is that the computer-generated images are the most convincing when they respect the tactical principles later developed during real conflicts, for example in the Korean War. Are these elements sufficient to make a scenario of this type a thought experiment?

This question would never have been asked fifty years ago. But during the last two decades of the 20th century, a wave of thought experiments hit all sorts of areas of knowledge; in particular philosophy. Highly imaginative examples, seemingly coming straight out of a B movie, flourished in contemporary philosophy: Searle's Chinese Room, Jackson's Mary and Dennett's Mary, Putnam's Twin Earth, variations on the Brain-in-a-vat, Thomson's violinist, and sundry zombies are the best known of this zoo. Historians of science and philosophy saw no reason to stay on the sidelines, and claimed the title of thought experiment for almost

[1] I would like to thank Christoph Lehner for discussing the issue of Einstein's thought experiments with me, and Carla Rita Palmerino for her comments on a previous version of this paper.

any argument: Hobbes' and Locke's states of nature, Carneades' plank, Descartes' evil genius, Hume's hypotheses on the annihilation of gold or paper money, Theon and Dion, Molyneux's problem, Buridan's ass, Boyle's mechanical angel, Archytas' man standing at the edge of the universe, Condillac's statue, Zeno's paradox, Leibniz' mill, Gyges' ring, Huygens' boat, the ship of Theseus.[2] Entire books are now devoted to collecting philosophical thought experiments of this kind.[3] Last, but not least, the field of science fiction, not to the say the broader field of literary fiction, came to be seen in general as a repository of thought experiments.[4]

Along with this came debates and attempts of philosophical clarification. In one of his seminal papers, Thomas Kuhn pointed out the paradox of scientific thought experiments: since they rely exclusively on familiar data, how can they lead to new knowledge of nature?[5] It was only in the early 1990s that a systematic debate began concerning the epistemological status and cognitive functioning of thought experiments.[6] The first question was epistemological, to decide if, as James Robert Brown argued, thought experiments are glimpses in a Platonic world of eternal laws, or if instead they are arguments relying on previous experiences, as the empiricist John Norton argued.[7] Another related but nevertheless distinct question concerned the kind of knowledge involved in thought experiments: to say, like Norton, that thought experiments are arguments which may imply (and indeed implied for Norton) that there are only verbal inferences disguised in vivid and picturesque narratives. Against what could be called "the inferentialist position" of Nor-

---

[2] Ierodiakonou, in "Ancient Thought Experiments," established a list of arguments that are now considered as thought experiments, although they were not conceived as such by ancient philosophers; the same could be done for other periods as well.

[3] Cohen, *Wittgenstein's Beetle*; Tittle, *What if* ....

[4] This trend can take inspiration from Ursula Le Guin, who analysed explicitly her own novel *The Left Hand of Darkness* as a thought experiment on gender; see Le Guin, "Introduction," 156; eadem, "Is Gender Necessary?," 163–167.

[5] Kuhn, "Function for Thought Experiment," 241; Horowitz and Massey, *Thought Experiments in Science*, 1.

[6] Horowitz and Massey, *Thought Experiments in Science*; Sorensen, *Thought Experiments*; idem, "Thought Experiments and the Epistemology."

[7] Brown, *Laboratory*; idem, "Platonic Account"; idem, "Thought Experiments Transcend"; Norton, "Thought Experiments in Einstein"; idem, "Are Thought Experiments"; idem, "Why Thought Experiments." In their contribution to this volume, Goffi and Roux start from the weak thesis that thought experiments are arguments, to be distinguished from the strong Nortonian thesis that thought experiments are *only* arguments.

ton, different non-inferentialist positions were formulated, according to which thought experiments cannot be reduced to arguments and inferences from one proposition to another, because they involve other cognitive mechanisms. Tamar Szabó Gendler explicitly defended an imaginarist position, but the non-inferentialism was already implied by the references to the notions of embodied knowledge (David Gooding) or mental models (Nancy Nersessian).[8]

As for Kathleen Wilkes, she cast doubts on the use of highly counterfactual thought experiments in ethics, most particularly concerning the questions of personal identity. An important point in her argument was the distinction between scientific and philosophical thoughts experiments. According to her, scientific thought experiments describe adequately their background, deal with natural kinds, and rely on impossible assumptions that are not relevant for the conclusion, while philosophical thought experiments give an inadequate description of the background, deal with indeterminate notions, and rely on impossible assumptions that are directly relevant for the conclusion.[9] A debate on the eventual distinction between scientific and philosophical thought experiments was thus opened, the distinction being generally made in order to condemn speculative philosophical thought experiments.[10]

The purpose of this introduction is not to answer each of the questions that may have been asked these past twenty years. Most of these questions are touched upon or even answered in the contributions of this volume. Instead, it is to raise the curiosity of readers by asking them to step back from the trivial use of thought experiments that has now become the rule. The essays gathered in this book are not concerned with contemporary scientific thought experiments. However, it is in contemporary science that the notion of thought experiments was first developed; nowadays, scientific thought experiments are supposed to work, and thus to constitute the standard according to which thought

[8] Gooding, "What is Experimental"; Gendler, "Galileo"; idem, "Thought Experiments Rethought"; Nersessian, "In the Theoretician's Laboratory." In his contribution to this volume, Zeimbekis puts to the test the thesis that thought experiments are simulations.

[9] Wilkes, *Real People*.

[10] Rescher, *What if?*. In asking the question of how to distinguish thought experiments that work and thought experiments that do not work, Atkinson and Peijnenburg, "Thought Experiments Poor," initiated a salutary change. In his contribution to this volume, Engel discusses the possibility of philosophical thought experiments.

experiments in other domains are judged. That is the reason why I shall begin by exploring the texts in which the origins of the scientific notion of thought experiments are usually said to be found. My general claim is simple: the emergence of the notion of thought experiments relies on a succession of misunderstandings and omissions. I shall then examine, in a more systematic perspective, the three characteristics of the broad category of thought experiments nowadays in circulation: thought experiments are counterfactual, they involve a concrete scenario and they have a well-delimited cognitive intention. My aim in exploring these characteristics is twofold. Firstly, it is to show that each of these characteristics, considered individually, may be taken in a more or less strict sense, and consequently to explain the proliferation of thought experiments. Secondly, it is to suggest that the recent debates on thought experiments might have arisen because these three characteristics are not easily conciliated when they are considered together. Finally, in a third and last section, the nine essays of this book will be presented.

1. *The Confused Origins of the Scientific Notion of Thought Experiments*

Depending on whether it is the term, the epistemological description, or the scientific practice of thought experiments that is considered, the origins of the notion of thought experiments are usually traced to Hans Christian Ørsted, to Ernst Mach, or to Albert Einstein. In this narrative, Ørsted created the term; Mach ennobled the notion by making it a specific scientific procedure; Einstein invented scientific thought experiments that continue to nourish our imagination. As often, these alleged origins are not proper explanations: it was only once "thought experiment" prevailed as a catchword that we began reading Ørsted, Mach, and Einstein in this light. An examination of their works is nonetheless instructive to understand the misunderstandings and omissions on which the notion of thought experiments was constituted. The term "thought experiment" is indeed to be found in Ørsted, but not in relation to a well-defined procedure; contrary to what is sometimes written, Mach never allowed for counterfactual thought experiments; lastly, Einstein certainly practised thought experimenting, but it was in scientific and epistemological circumstances so specific that it might be doubted if his practice might be an epistemological warrant for thought experiments in general.

1.1. Lichtenberg sometimes invoked the possibility of "experimenting with thoughts," especially when reasoning on particular cases.[11] However, the first to use the term "thought experiment" (*Gedankenexperiment, Gedankenversuch*) in a scientific context may have been Hans Christian Ørsted.[12] In his "First Introduction to General Physics" (1811), he defended, against the excessive speculations of the *Naturphilosophie*, the thesis that the natural sciences cannot dispense with experience, whether everyday experiences, observations, or scientific experiments.[13] However in that context he insists that there is a higher level of the experimental art that sets the mind into "creative activity" (§§ 16–17). This kind of activity manifests itself not only in the natural sciences, but in mathematics as well, in so far as mathematical objects may be engendered by the mind. It is at this point that thought experiments enter the scene. Allowing a point to move in order to draw a line, or using certain procedures of the differential and integral calculus would be "thought experiments" that do not only show that something is in a certain way, but why it is in this way. According to Ørsted, a similar genetic method should be introduced in physics; on that occasion, he refers to Kant's *Metaphysical Foundations of Natural Science* as the book that includes "the most beautiful examples of this kind of presentation."[14] The mention of Kant may indicate that thought experimentation would amount to constructing concepts in intuition. In any case, Ørsted in the end proposes to bring together mathematics and physics by presenting both as "a series of thought experiments."[15]

The next paragraphs (§§ 18–19), while continuing to speak of thought experiments, present nonetheless considerations of a totally different order.[16] Here, thought experiments are no longer tied to the genetic

---

[11] Schildknecht, *Philosophische Masken*, 147 sqq.; Kühne, *Gedankenexperiment*, 220–221.

[12] As was first pointed out by Witt-Hansen 1976; see further Kühne, *Gedankenexperiment*, 130–165; Cohnitz, "Ørsteds Gedankenexperiment." Kühne, *Gedankenexperiment*, 151–161, examines as well some counterfactual situations that Ørsted calls "thought experiment" in his scientific writings.

[13] Ørsted, "Introduction," 289, 292–293. The English edition that I am following here is actually a translation of the 1822 version of Ørsted's essay, published in German in the *Gehlens Journal für Chemie und Physik*.

[14] Ørsted, "Introduction," 296–297.

[15] For tentative interpretations of these obscure lines, with references to the mathematics and philosophy of mathematics of the early 18th century, see Kühne, *Gedankenexperiment*, 138–147; Cohnitz, "Ørsteds Gedankenexperiment."

[16] Kühne, *Gedankenexperiment*, 137–138, 147, notes as well this conceptual discontinuity.

method that should be used to reform mathematics and physics, but are assimilated to conjectures made in the context of the hypothetico-deductive natural sciences, that begin with the assumption of a hypothesis, continue with the examination of its consequences, and end up in confrontation to actual experiments. Since we are not able to truly confirm a hypothesis by checking that all the consequences that can be deduced from it actually occur, Ørsted finally notes that a hypothesis can never be certain and describes some procedures that would increase its probability.[17]

It should be concluded that, for Ørsted, thought experiments are not a specific epistemological procedure: rather, they are in a more general fashion the workings of thought itself, in so far as it is able to move from one proposition to another. Indeed, the point in common between what Ørsted calls a thought experiment in mathematics (§§ 16–17) and a thought experiment in physics (§§ 18–19) seems to be that in each of them one is led to follow a certain train of thought when examining the consequences that follow, either from the construction of a mathematical object, or from the position of a hypothesis in physics. Thus, it seems difficult to say that there is a specific and well-delimited notion of thought experiments in Ørsted. As I shall now show, the situation improves a bit when we turn to Ernst Mach; he was indeed the first to construct explicitly a notion of thought experiment, but, I shall argue, it does not include counterfactuality.[18]

1.2. In *The Science of Mechanics* (1893), Mach discusses Stevin's necklace, still considered one of the canonical examples of thought experiments. Stevin's necklace has a well-delimited cognitive intention, to prove the law of the inclined plane, stating that the ratio between a weight and the force needed to balance this weight on a given inclined plane is equal to the ratio between the length and the height of this plane—that is to 1/sin*a*, *a* being the angle of inclination. As a thought experimenter, who wants to prove this law in his armchair, Stevin asks you to consider the following concrete scenario. You have a triangle whose sides are inclined planes of the same height and whose hypotenuse is parallel

[17] Ørsted, "Introduction," 297–299. For latter versions of the same argument (1851), see Kühne, *Gedankenexperiment*, 149.

[18] There is no historical evidence that Mach read Ørsted's essay; nevertheless, see Kühne, *Gedankenexperiment*, 166–167, 187–188. For alternative discussions of Mach, see Sorensen, *Thought Experiments*, 51–75; Kühne, *Gedankenexperiment*, 167–214.

to the horizon, and you surround it with a necklace of identical and evenly spaced beads. When you ask yourself what the "power" of one of the beads on an inclined plane is, that is, the vertical force needed to balance its weight, your reasoning is as follows: Mechanical perpetual motion being impossible, you know that the necklace is in equilibrium. You know that if you remove equal things from equal things, you have equal things; hence if you eliminate the part of the necklace that is under the inclined plane, the remaining parts on the top and the slope are still in equilibrium. But, by construction, the number of beads on a plane is proportional to its length. You finally conclude that the power of a bead on a plane is inversely proportional to the length of this plane.[19]

Commenting on this scenario, Mach asks the unavoidable question for an empiricist: how is it that its result seems to us more certain than the result of a real experiment or than the outcome of a deduction, and that there is a generality in it that goes beyond the particular set-up imagined? His answer is that a thought experiment taps into a store of instinctive knowledge and summons up beliefs from this store with respect to a specific problem:

> Everything which we observe imprints itself uncomprehended and unanalysed in our percepts and ideas, which then, in their turn, mimic the process of nature in their most general and most striking features. In these accumulated experiences we possess a treasure-store which is ever close at hand and of which only the smallest portion is embodied in clear articulate thought. The circumstance that it is far easier to resort to these experiences than it is to nature herself, and that they are, notwithstanding this, free, in the sense indicated, from all subjectivity, invests them with high value.[20]

The idea that mechanical science relies on an instinctive knowledge that we build up thanks to our common experience of the things of the world and owing to more elaborate practices involving our artificial machines, is a main idea in *The Science of Mechanics*.[21] What differentiates this instinctive knowledge from scientific knowledge is that it is neither conscious nor reflected; the very use of the term "instinct" shows

---

[19] *De Beghinselen der Weeghconst* (1586), I, theor. 11, prop. 19, analysed in Mach, *Science of Mechanics*, 32–34. As for the term "thought experiment," Mach introduces it somewhat belatedly: "It is not essential that the experiments should have been actually carried out, if the result is beyond doubt. As a matter of fact, Stevinus experiments mentally" (ibid., 38). And this is no systematic use: he speaks as well of Stevin's "deduction" or "fiction" (ibid., 35, 601–602).

[20] Ibid., 36.

[21] Ibid., 1–5, 103.

that Mach casts his reflections in an evolutionary framework where humans and animals only differ in degree, and where certain truths, for example the principle of causality, are acquired by "the development of the race."[22] Inasmuch as instinctive knowledge is unconscious and pre-reflected knowledge, it is, as it was, the spokesman of nature itself, "free of all subjectivity."[23] This in turn explains why it enjoys our exceptional confidence, even though it is subject to error.[24] But, as a good empiricist, Mach insists that we should not be compelled to create a new mysticism out of this instinctive knowledge: it does not come from sources other than past experiences.[25]

Mach's empiricism has serious consequences regarding thought experiments. When faced with Newton's bucket, Mach strictly marked off the epistemological limits of thought experiments. You may recall that Newton's bucket is supposed to prove the existence of absolute space: inasmuch as rotation has real effects (when the water in the bucket begins to rotate with the bucket, it does not have any motion relative to the bucket, but the surface of the water becomes concave), it should be measured with respect to absolute space. Mach's objection, first expressed ironically—"try to fix Newton's bucket and rotate the heaven of fixed stars and then prove the absence of centrifugal forces"—is an epistemological objection. Facing an offspring of the Bucket worked out by Carl Neumann, who, unlike Newton, explicitly assumed that the rest of the universe disappeared, Mach criticises him for having made

> too free a use of intellectual experiment .... When experimenting in thought it is permissible to modify *unimportant* circumstances in order to bring out new features in a given case; but it is not to be antecedently assumed that the universe is without influence on the phenomenon here is question.[26]

Newton's experiment is a real experiment indeed, but inferring from what happens to the bucket to the existence of an absolute space is clearly extending our principles beyond the boundaries of experience: Mach condemns such an extension as "meaningless,"[27] and calls absolute space

22 Ibid., 1, 581.

23 This is linked to Mach's sensationalism that appears briefly in ibid., 578–579, 611.

24 Ibid., 34–36, 91–92, 94.

25 Ibid., 27, 94. Kühne, *Gedankenexperiment*, 172–175, insists on the anti-metaphysical and anti-Kantian background of this assertion.

26 Mach, *Science of Mechanics*, 340–341.

27 Ibid., 279–280.

a "monstrous conception": monstrous, precisely because it is created by the mind without the safeguard of experience.[28]

The nature, the function, and the domain of thought experiments are pretty clearly delimited in *The Science of Mechanics*. If we perform thought experiments, this does not imply that they are of a different nature than real experiments and that we resort to an extra-sensory faculty: they are performed because we have unanalysed real experiences stored in our memory at our disposal—today we would rather say that they constitute our "implicit" or "tacit" knowledge. As for the function of thought experiments, it is to trigger, through a specific scenario, a mental process that allows us to mobilise our instinctive knowledge in order to develop new explicit knowledge. Lastly, the domain of thought experiments seems consequently restricted to the domain of elementary mechanics, in which we can rely on previous experiments we have had with bodies, raising them, pushing them, pulling them, sliding them, rotating them in order to invent specific scenarios.

In *Knowledge and Error* (1905), which resumed a paper first published in the *Zeitschrift für physicalischen und chemischen Unterricht* (1896), Mach did not give up his empiricism, but extended the limits of experiments so far that they become almost as loose and indeterminate as in Ørsted's *First Introduction to General Physics*. Here is how he actually introduces the notion:

> Besides physical experiments there are others that are extensively used at a higher intellectual level, namely thought experiments. The planner, the builder of castles in the air, the novelist, the author of social and technological utopias is experimenting with thoughts; so, too, is the hard-headed merchant, the serious inventor and the enquirer. All of them imagine conditions, and connect with them their expectations and surmise of certain consequences: they gain a thought experiment. However, while the former combine in fantasy certain conditions that never occur together in reality, or imagine these conditions accompanied by consequences that are not connected with them, the latter, whose ideas are good representations of the facts, will keep fairly close to reality in their thinking. Indeed, it is the more or less non-arbitrary representation of facts in our ideas that make thought experiments possible.[29]

According to this text, thought experiments in general result from a basic tendency of the mind to imagine conditions resulting from the combination of ideas with one another and to examine their consequences: this is

[28] Ibid., Preface to the 7th edition: xxviii.
[29] Mach, *Knowledge*, 136.

what Mach called the mutual adaptation of thoughts. However, there is a species of thought experiments that deserves a special mention because it implies as well an adaptation of thoughts to facts, namely, thought experiments that unlike the fanciful scenarios of dreamers and novelists, combine "ideas that are good representations of the facts."[30] In that sense, true thought experiments are so to speak loaded with experience upstream. They also have connections with experience downstream. Two kinds of thought experiments are indeed distinguished by Mach: those very few that by themselves can be carried to a definite conclusion (as was the case with Stevin's necklace) and those which, lacking a definite outcome, call for physical experiments.[31] In the latter case, thought experiments "precede and prepare" physical experiments; they are mental anticipations that help us conceive experiments properly and bring them into being.[32] According to Mach, these mental anticipations constitute the true *ars inveniendi* in the sciences, but it is only an *ars inveniendi*, and by no means an *ars demonstrandi*. For example, Galileo knew that speed increases in free fall and made a guess concerning this increase; this guess allowed him to design the proper experiment; if he had not performed the experiment, it would have remained a conjecture in want of a proof.[33] Hence, as in *The Science of Mechanics*, though via different paths, Mach tied up thought experiments and physical experiments in *Knowledge and Error*.

For a comprehensive outline it should finally be noted that this is in a sense counterbalanced by the end of the chapter, where Mach insisted on the method of continuous variation. His idea was if adaptation of thoughts to each other and to facts emerges from a biological need, science should help this adaptation by systematic procedures. The method of continuous variation, which consists in imagining how known facts would vary when one of the relevant parameters varies, as much as possible in a continuous manner, would be one of these procedures.[34] Inasmuch as this method is used in mathematics, Mach extended to mathematics the use of thought experiments; finally, he ended up blurring

---

[30] On the distinction between free fiction and non-arbitrary science, see ibid. p. 137: "Since the physicist always turns his thoughts towards reality, his activity differs from free fiction."

[31] Ibid., 136, 138, 142, 148. Stevin's necklace is actually mentioned in the previous chapter as an example of mutual adaptation of thoughts (126).

[32] Ibid., 136, 138, 141, 148.

[33] Ibid., 142. This example is developed in Mach, *Science of Mechanics*, 158–163.

[34] Mach, *Knowledge*, 137–138, 149.

the outlines between experiments and deductive thought.[35] This should not surprise us: in so far as the notion of thought experiments promoted in *Knowledge and Error* is linked to the biological need of adaptation of thoughts and to the reflexive method of variation, it tends to lose all reference to specific set-ups and to be assimilated to general epistemological processes. Hence my conclusion on Mach's notion of thought experiments is twofold. Firstly, in *Knowledge and Error*, there is a tendency, as was already the case in Ørsted's *First Introduction to General Physics*, to dilute the notion of thought experiments in other more general epistemological processes. Secondly, in *The Science of Mechanics*, as in *Knowledge and Error*, Mach stressed that thought experiments are no substitute for physical experiments; they always presuppose well-established facts; in most cases, they simply offer hypothetical anticipations of physical experiments to come. According to Mach, highly counterfactual thought experiments are of no use in physics.

1.3. Every elementary textbook on thought experiments begins with or at least includes a catalogue of Einstein's famous thought experiments: Einstein's trains, Einstein's lift, Einstein's chase after a beam of light ... However, Einstein himself did not use the term "thought experiment" for what is now referred to as "Einstein's thought experiment": with variations from one text to the other, he spoke of "example," "argument," "analogy," "illustration," "idealised experiment" etc. He opened, for example, his special relativity paper of 1905 by focusing on the case of the electro-dynamic interaction between a magnet and a conductor in relative motions. This is the first example discussed in Norton's seminal paper on thought experiments in Einstein's works, but for Einstein himself, this was not a thought experiment, but an "example" of the principle of special relativity which he introduced at this point as a "conjecture."[36] As we will see and comment in some detail, in the popular book he wrote with Leopold Infeld, the law of inertia, the lift thought experiment, and an electrical circuit reduced to a point were called "idealised thought

[35] Ibid., 144–146.

[36] Norton, "Thought Experiments in Einstein," 135–136, and Einstein, "On the Electrodynamics," 140. Einstein and Infeld, *Evolution of Physics*, 125–128, makes clear that this example relies on a comparison between two well-known real experiments, the one by Ørsted and the other by Faraday. Hence, it is a thought experiment only in so far as we admit thought experiments with a weak counterfactuality (on weak, average, and strong counterfactuality, see more below).

experiments."[37] And if we eventually jump to the end of Einstein's career, in his *Autobiographical Notes*, the mechanics of people who might know only a small part of the earth and who would not see any star at all is an "analogy" to let us understand Mach's critique of mechanics; his chasing a beam of light at the age of sixteen is a "paradox"; lastly, the so-called EPR thought experiment is a "reasoning."[38] The absence of the term "thought experiment" cannot be explained by Einstein's ignorance: on the contrary, it is well-known that Einstein read Mach's *Science of Mechanics* in his earlier years and that he drew inspiration from its discussions on relativity.[39] And if we look hard, we can even find a few scenarios that Einstein did indeed call "thought experiments."[40]

These facts may raise a doubt: if Einstein chose as a rule not to use the term "thought experiment," even though he was perfectly familiar with it, we can wonder if this is a pertinent epistemological notion to analyse his works. And yet, beside the fact that this doubt would lead to a very drastic reduction in our use of the category of thought experiments, the Ørsted episode shows that we cannot let ourselves be guided by words alone. In this case, even if Einstein barely used the term "thought experiment," he did employ specific methods of argumentation, which we can designate, in accordance with the now well-established usage, as thought experiments. In order for this to be totally legitimate, we must preliminary explain that as a rule he avoided the term "thought experiment." An episode from the first reception of the theory of relativity will allow me to answer this preliminary question.

In a popular presentation of relativity (1917), Einstein had introduced what are now called "the trains/embankments thought experiments," which he himself presented as "illustrations" to help his intended readership grasp the essentials of the theory of relativity without being obliged to cope with the mathematical apparatus of theoretical physics.[41] Philipp Lenard, in *Über Relativitätsprinzip, Äther, Gravitation* (1918), turned Einstein's arms against him by inventing his own scenarios intended to

[37] Einstein and Infeld, *Evolution of Physics*, 6–8, 144, 214.

[38] Einstein, *Autobiographical Notes*, 27, 49–51, 79–81. For the same conclusions and other examples, see Kühne, *Gedankenexperiment*, 227–228.

[39] Kühne, *Gedankenexperiment*, 234–236, assumes that Einstein could not ignore Mach's notion; for an account of Mach's and Einstein's relations, see Holton, "Mach, Einstein."

[40] Kühne, *Gedankenexperiment*, 236–244, examines them and holds that they are insufficiently well determined scenarios to give rise to experiments, or a strategy of mockery of objections that are insufficiently physically grounded.

[41] Einstein, "A Popular Account," 310 and 248 for the intended readership.

establish the counter-intuitive character of general relativity. If a train suddenly brakes near a church, he argued for example, objects in the train are thrown about; it would hurt the "plain common sense, that could also be called healthy" to attribute these effects to a change of motion of the nearby church. In a discussion with Lenard that occured at the congress of the *Gesellschaft Deutscher Naturforscher und Ärzte* in Bad Nauheim (1920), Einstein stressed that the criteria of intuitiveness (*Anschaulichkeit*) as understood by Lenard could not be used: physics is not intuitive, but conceptual; moreover, the Galilean episode shows that intuitiveness is historically variable. In particular, although he could have retracted by arguing that trains and embankments were only illustrations introduced in a popular book, he fully accepted their use and designated them as thought experiments:

> A thought experiment is an experiment that can be carried out in principle (*ein prinzipiell ausführbares Experiment*), even if not in fact (*faktisch*). It is used to give a clear view of a set of genuine experiments (*wirkliche Erfahrungen*), so as to draw the theoretical consequences of them. A thought experiment is only forbidden if it is impossible to be carried out in principle (*prinzipiell unmöglich*).[42]

In this discussion at least, Einstein used the term "thought experiment" in reference to his own tools of argumentation. We also find a plausible explanation why he did not use the term often. As we have said, Einstein knew *The Science of Mechanics* perfectly well, in particular the passages in which Mach, defending his own principle of relativity, declares that thought experiments not guided by past and future physical experiments are meaningless. Yet one of the most striking characteristics of Einstein's thought experiments, that he defended here, is that they are counterfactual. Of course, he noted that not just any counterfactuality will do, since a thought experiment must "be possible to carry out in principle." But he did not tell us—and I think because it is extremely difficult, not to say impossible—where the line runs between what can be carried out in principle and what cannot.[43] In other words, my conjecture is that Einstein generally avoided the term "thought experiment" because he was aware

---

[42] Einstein, "Lectures in Bad Nauheim," 354–355, my translation; as noted by the editors (109), the transcription of these discussions is incomplete. On the historical context of the Bad Nauheim encounter and Lenard's anti-Semitism, see idem, 101–113. On Lenard's subsequent positions on thought experiments in the context of his Aryan physics, see Kühne, *Gedankenexperiment*, 259–260.

[43] In his contribution to this volume, Virvidakis suggests that this line may be drawn following Kant's analysis of transcendental conditions of experience; but even if Einstein

that in introducing counterfactuality into physics, he set himself against Mach's legacy without having a clear procedure for distinguishing acceptable and inacceptable counterfactuality.

Now that the preliminary question of why Einstein avoided the term "thought experiment" has been answered, I would like to go further by concentrating on one of the most famous Einsteinian thought experiments, the lift thought experiment. Of course the lift is not the only thought experiment created by Einstein, and Einstein was not the only physicist who proposed thought experiments in those years. Just think, for instance, of the series of thought experiments linked to the interpretation of quantum mechanics that would peak in the EPR paradox, or of Heisenberg's microscope and Schrödinger's cat. A detailed discussion would be needed for each of them, if we wanted to grasp the epistemological function of each of them. But the discussion of the lift thought experiment will be sufficient to enlighten the counterfactuality of Einstein's thought experiments, and to relate it, in this case, to its epistemological context.

Einstein's lift is intended to defend the extension of this principle of special relativity to the case of accelerated motion, in other words, the equivalence of a homogeneous gravitational field with a uniformly accelerated frame of reference. It consists in imagining, in an area devoid of gravity, a closed opaque box, which is pulled with a constant force, so that it has an accelerated motion. An observer inside this box feels a pressure in his legs and sees all the bodies in the box falling with the same acceleration. This observer, having consequently no way of distinguishing his situation from the situation of a man in a box at rest in a gravitational field, would have good reason to think that he is in a gravitational field. Some details differ in the various presentations of this thought experiment: in one of them, there are actually two physicists inside the box, who happen to be there because they were put to narcotic sleep; the box is an opaque chest or a great lift of a skyscraper much higher than any real one; the objects that fall in the box may be either unspecified objects or a handkerchief and a watch.[44]

were a Kantian, the distinction between logical possibility and transcendental possibility would not help him to decide if a given experiment is likely to be carried out in principle.

[44] Einstein, "Present State of Gravitation," 208; idem, "A Popular Account," 318–320. Einstein and Infeld, *Evolution of Physics*, 214; the "idealised reader" is depicted in the Preface: "We had him making up for a complete lack of concrete knowledge of physics and mathematics by a great number of virtues" (n. p.).

It is in papers intended for a broader readership that this thought experiment was presented, while, as noted by Norton, in more scientific papers the argument appeared, so to speak, naked, without the thought experiment.[45] Hence, it cannot be doubted that, as Einstein's trains and embankments, Einstein's lift had an illustrative and didactic function. He resorted to it in popular writings, in which he avoided mathematical formalism as well as physical technicalities: it was a striking illustration that put its seal on a theory that was otherwise well established, without in and of itself being what establishes this theory. But while it has this illustrative and didactic function, it is not the only function it has. Based on the few explicit declarations from Einstein about his epistemology, we can in fact show that it also has a heuristic function: it guides us in the formulation of new principles and allows for the constitution of new forms of intuitiveness.

In the work, in which he is the most explicit regarding the status of his lift thought experiment, Einstein describes it as an "idealised experiment," similar to the one Galileo used to formulate the founding statement of modern physics, the law of inertia. In order to formulate this law, Galileo had suggested against intuitive conclusions based on immediate observations to contemplate a body moving forever with no friction nor any other external forces acting on it; in the same way, it would be legitimate to propose experiments in which lifts are pulled by immaterial beings in radical voids devoid of any gravitational field. The two texts—the text in which Einstein explained how Galileo invented the law of inertia and the text in which he justified his own use of thought experiments—are quite coherent. According to them, "idealised experiments" present the following characteristics:

- they are "fantastic," they can neither be derived from experiment nor actually performed;
- they are consequently created "by thought" and even by "speculative thinking";
- they are however "consistent with observation" and lead to "a profound understanding of real experiments."[46]

[45] Norton, "Thought Experiments in Einstein," 138, referring to Einstein, "Influence of Gravitation," 379–380 and idem, "Tetrode and Sackur," 150.

[46] Einstein and Infeld, *Evolution of Physics*, 214, 6–8. As I will explain in the next part of this introduction, the fact that Einstein described these thought experiments as idealised experiments led to confusion, that should, however, be avoided.

In what follows, I will show that this description is consistent with what we know more generally about Einstein's epistemology, but that considering that the law of inertia and the lift fall under the same description irons out one decisive difference between them.[47]

The young Einstein was certainly an empiricist, as indicates his reaction to a letter in which his friend Michele Besso explained that in the theory of relativity speculation had revealed itself to be superior to empiricism (1918). In his answer, Einstein insisted that a theory "must be built upon generalisable facts," even adding that "no genuinely useful and profound theory has ever really been found purely speculatively." But what Einstein called "facts" in this letter might already be a surprise: the impossibility of *perpetuum mobile*, the law of inertia, the equivalence of heat and mechanical energy, the constancy of light velocity.[48] However, only gradually did he come to realise how much his own epistemology differed from classical empiricism, in particular because he thought that all concepts, even those apparently closest to experience, are not derived from experience, but freely chosen and constructed, provided that they are not in contradiction with experience.[49]

His position was adequately described in a letter to Solovine (1952). To make his epistemological position understood, Einstein made a diagram showing three levels: from top to bottom he showed the axioms (A); the statements deduced from them (S); the variety of immediate experiences (E). According to Einstein, the only logical connection is the connection between the As and the Ss: the two other relations—the relation of the Ss to the Es and the relation of the Es to the As—are qualified as "intuitive," "psychological," and "extra-logical."[50] We can thus understand the function of idealised experiments such as the lift thought experiment according to Einstein. They are devices that allow us not to attribute any form of necessity to the relationship between Es and As, but rather to heuristically guide the intuition, which, based on the variety of experiences and observations (Es), identifies the principles (As) that can be taken to be axioms in the deductive system of theoretical physics. Note moreover that the As

[47] Einstein's epistemology is extensively discussed in Einsteinian scholarship, in particular to determine whether it evolved or not, and whether Einstein should or should not be categorised as a realist. See in particular Holton 1970; Fine, *The Shaky Games*; Brown, *Laboratory*, 99–125. These issues do not need to be discussed here.

[48] Einstein to Michele Besso, 28 August 1918, in Einstein, *CPAE*, Vol. 8, 633; see further ibid., 638.

[49] Einstein, *Autobiographical Notes*, 13, 19–21.

[50] Einstein to Solovine, 7 May 1952, in Einstein, *Letters*, 137–138.

are such general principles that, once we have grasped them, we are able to deal with counterfactual situations that we have never experienced. This is what is at issue with Einstein's lift: its goal is to go back from the experience we have of a gravitational field to the principles that allow us to imagine it in the framework of a new theoretical physics—here, laws of gravitation formulated for all coordinate systems, not only for inertial coordinate systems.

This function of thought experiments is decisive when a new principle theory is introduced, as was the case with the theory of relativity. Remember that Einstein considered the theory of relativity as a principle theory, not as a constructive theory. The difference between these two types of theories according to him is the following. Constructive theories build up a picture of more complex phenomena out of a simple theoretical hypothesis: for example the kinetic theory of gases builds the mechanical, thermal and diffusional processes out of the hypothesis of molecular motion. In contrast to this, the starting point of principle theories are "empirically observed general properties of natural phenomena, principles from which mathematical *formulae* are deduced of such kind that they apply to every case that presents itself": we recognise here the "facts" of the letter to Besso, and, indeed, Einstein gave the example of the science of thermodynamics that starts from the fact that perpetual motion is impossible.[51] But a problem arises when phenomena are already subsumed under other principles: what kind of empirical reference can justify the change from one set of principles to the other? The lift thought experiment is a kind of justification, that proceeds by constructing a simili-fact that makes us simili-experience the possibility of shifting from the old set of principles to the new set of principles. This explains the profound affinity between Galileo and Einstein: for both of them, thought experiments allow the introduction of new overarching principles of great comprehensiveness, different from those previously admitted. Considering their great comprehensiveness, these principles cannot be derived from or confirmed by real experiences; but it is possible to use thought experiences as guides for elaborating intuitions. In this respect, Einstein's lift seems the exact equivalent of Galileo's cabin, in which butterflies fly, fish swim and drops of water fall exactly in the same way whether the ship is, relative to the earth, standing at rest or moving

[51] Einstein, "Time, Space and Gravitation," 213; Brown, *Laboratory*, 103–117, uses differently the distinction between the two types of theories.

inertially. In both cases, there is a situation that can be interpreted according to two different frameworks.

Nonetheless, although Einstein assimilated his approach with that of Galileo, it seems inevitable to distinguish them. What triggers Galileo's cabin is the fact that we live two types of experiences which cannot be distinguished as experiences: the experience of being in the cabin of a ship that stands still, and the experience of being in the cabin of a ship sailing on a calm sea. It is because of these two types of experiences that the theoretical equivalence of being at rest and being in inertial motion is experientially convincing; and, of course, once we have established this equivalence, the reasoning can be extended to the earth. In contrast, despite the vivid details offered by Einstein, we have no occasion to compare experiences that would convince us that being in a gravitational field and being in Einstein's lift are equivalent; we have only *one* lived experience, the experience of being in what we call a gravitational field. Hence, Einstein's point cannot be that two distinct experiences should be seen as one single theoretical case, but on the contrary that one single experience can be interpreted according to two distinct theoretical frameworks. In this respect, it is not immaterial that the human beings in the lift are physicists, who not only live experiences, but formulate physical theories in a mathematical apparatus. To put this in a nutshell, Galileo's ship and Einstein's lift share many resemblances: both are associated with a radical change in the principles of mechanics; both are material set-ups designed for the transportation of things and living beings; both result in presenting two elements as equivalent. But the elements presented as equivalent are not the same: in the case of Galileo's ship, they are two experiences; in the case of Einstein's lift, they are two possible theoretical views (either a gravitational field or an accelerated frame of reference) of one and the same experience (being in a terrestrial lift). In other words, even though Einstein insisted that his approach is similar to Galileo's, his thought experiment has counterfactuality that has no equivalent in Galileo.[52] In this respect, Einstein's lift is exemplary enough to reveal a general characteristic of thought experiments, as we now understand them: not only do they have a well-determined cognitive intention and involve a concrete scenario, but also they are counterfactual. This third characteristic emerged in Einstein physics, and the intellectual authority

[52] That Galilean thought experiments are much more restricted in scope than sometimes described, is amply established in Palmerino's contribution to this volume.

he held allowed for the crystallisation and spreading of a category of thought experiments fashioned from these three characteristics. Now, I would like to examine this category in a more systematic fashion.

2. *The Extensible Category of Thought Experiments*

As I have said earlier, if we try to determine what thought experiments are, we will most probably end up with the following three characteristics: they are counterfactual, they involve a concrete scenario, and they have a well-determined cognitive intention. Like any characterisation, this one raises two questions. Firstly, the question of its empirical coverage: does it truly include the cases usually described as thought experiments? And secondly, the question of its coherence: do its three characteristics not present certain tensions, if not contradictions as such? The first question has been regularly asked as a preliminary question, with the objective to indicate in advance the limits of a definition that would then be given.[53] I shall approach this question differently by showing that the category of thought experiments is more or less inclusive, depending on the leeway with which we take each of its characteristics. The second question, however, has never been asked. Therefore I would also like to suggest that bringing together these three characteristics, especially if some leeway in each of them is allowed, might lead to tensions.

2.1. Like the model-maker scenario that opens this introduction, as opposed to real experiments, a thought experiment does not have to take place in reality. We can reach its result merely by thinking: it is, as we say in philosophical parlance, counterfactual. It is in relation to this characteristic that we hear about spectacular thought experiments that deal with situations, which we believe were never carried out, or which may be altogether impossible to carry out. Between zero counterfactuality, which corresponds to reality, and maximal counterfactuality, which deals with metaphysical and logical impossibilities, it seems useful to distinguish among various degrees of counterfactuality at this point.[54] I would

[53] See for example Kuhn, "Function for Thought Experiment," 241; Brown, "Platonic Account," 122; Nersessian, "In the Theoretician's Laboratory," 295.

[54] Albeit the distinction between weak and strong counterfactuality is broached in other words in Wilkes, *Real People*, 3; Brown, *Laboratory*, 36. Brown, however, does neither include Stevin's necklace nor Galileo's ship in his category of "merely imagined thought experiments."

especially like to suggest that there is a distinction between weak, average and strong counterfactuality:

i. Weak counterfactuality concerns thought experiments that are physically possible and that could have been produced in reality considering human ability and technical means available at the time they were proposed. It just happens that they were not carried out, for example because their actual implementation appeared redundant regarding what the person proposing them accepted as common beliefs and shared knowledge; or too demanding in terms of time, money, and other resources, etc. Thus, there is no call to place a necklace of beads around an inclined plane to observe the law of the inclined plane; we do not need to build and take apart the ship of Theseus to grasp certain paradoxes tied to identity; it is unnecessary to lock yourself in a room and transcribe messages in Chinese to understand the objection advanced by Searle against the computationalists in the 1970s, and so on.[55]
ii. Average counterfactuality involves scenarios which not only did not happen, but which could not happen considering our beliefs concerning human capacity to intervene in the world. This is the case, for example, with Einstein's lift: we do not believe that we might construct a box in an outer space devoid of any gravitational field and impress a constant force on it. This scenario seems to deal not only with a contingently unrealised possibility, but also with a technical impossibility. Since our perception of what is technically possible changes over time, the demarcation between cases of weak counterfactuality and cases of average counterfactuality is relative. The Einstein-Podolsky-Rosen (EPR) argument nowadays is famous not only because it appeared in 1935 as a crucial argument to show that either quantum mechanics is complete, or we have to renounce the principle of separation, but also because, almost fifty years later, it was finally put to the test by Alain Aspect, Philippe Grangier and Gérard Rogier in a paper suggestively entitled "Experimental Realisation of Einstein-Podolsky-Rosen-Bohm *Gedankenexperiment*: A New Violation of Bell's Inequalities."
iii. Lastly, strong counterfactuality concerns cases that we judge impossible with respect to the laws of nature of our world or even to our

[55] Thought experiments involving weak counterfactuality are examined in the Lautner's contribution to this volume.

> metaphysical tenets. In the early 17th century, not only were the technical means to produce a physical void not available, but some natural philosophers thought that it was a metaphysical impossibility, with the argument that a void is nothing and that attributing existence to nothing amounts to a contradiction. A thought experiment involving a void for them was consequently a case of strong counterfactuality, and they could argue against it that it is of no use when the point is to identify the behaviour of bodies in our real world.[56]

This distinction between three types of counterfactuality should be granted in principle. The only question is, how far do we want to extend weak counterfactuality. Perhaps because of a taste for the exotic, or to impress the layperson, contemporary literature usually highlights cases of strong counterfactuality.[57] Under these conditions, it is natural to counterbalance this trend and to recall that there are cases of weak counterfactuality. Consider however the following case of extrapolation: Galileo wanted to establish a law concerning bodies falling in a void, but he did not have the technical means to produce a void. He carried out experiments on bodies of various specific weights falling in mediums of greater or lesser resistance, observed that the inequality of the bodies' speed decreases when the resistance of the medium decreases, and so concluded by means of a controlled extrapolation that in a medium of zero resistance—a vacuum—the speed inequality is zero.[58] Should we include this extrapolation among thought experiments? If we do so, all extrapolations and idealisations should be included in the category of thought experiments. Physical experiments concerning ideal bodies, devised by isolating the relevant parameters in thought, mean we might end up with a physics peopled solely with thought experiments.[59]

---

[56] In their contribution to this volume, Knuuttila and Kukkonen present the conceptualisation that some medieval commentators proposed of thought experiments beginning with impossible premises.

[57] Against this tendency, see Brown, "Thought Experiments Transcend," 25; Rescher, *What if?*, 3–4.

[58] Galilei, *Two Sciences*, 75–76.

[59] The assimilation, not to say the confusion, between idealisation and thought experimenting is made in Koyré, "Le *De Motu gravium* de Galilée," 225; idem, "Pascal savant," 382–385; McAllister, "Evidential Significance," 245–268; idem, "Thought Experiments and the Belief," 1168–1170. Einstein's use of the term "idealised experiments" for thought experiments explains this confusion in part.

The following reflection may help us distinguish idealised experiments and thought experiments, even if the literature has conflated the two categories since Einstein. Thought experiments are performed "in the laboratory of the mind," as James Brown aptly said. Even if we do not concede to him that such a laboratory is closed to experience, his bit of metaphor is a good description of the fact that it is our thinking, which performs the experiment in thought experiments. By contrast, idealised experiments are experiments, which are in the first place performed out there in the real world, and only then integrated into our system of thoughts. This integration may indeed be described as an idealisation and it involves the introduction of some counterfactuality, but we cannot say that it is our thinking, which performs idealised experiments. Thus, a thought experiment is counterfactual because it is achieved in thought. Whereas introducing some counterfactuality in a real experiment does not make it necessarily a thought experiment.

2.2. A demonstration in mathematics is carried out in thought, and is often based on counterfactual hypotheses, which do not correspond to what exists in our world. For example, in the first book of *Principia mathematica*, Newton calculates the trajectory of bodies subjected to a law of attraction proportional to their distance, not that this law corresponds to anything real, but just because he enjoys developing the logical consequences of mathematical hypotheses.[60] But we certainly do not want to count this kind of mathematical counterfactual reasoning as thought experimenting; this would amount to seeing the whole of mathematics as resulting from thought experiments.[61] In order to exclude most mathematical reasoning from thought experiments, I need to avail myself to my second set of characteristics: unlike mathematical reasoning, thought experiments present vivid specific cases that may be called "scenarios." This characteristic is related to the fact that thought experiments are often associated with the names of their inventors, accompanied by those of their main protagonists, whether a person or an object (Leibniz' mill, Thomson's violinist, Stevin's necklace, Schrödinger's cat, Maxwell's demon, Jackson's Mary, Dennett's Mary etc.). Indeed, we find

[60] *Principia mathematica*, book I, proposition 44, problem 40.

[61] As I showed earlier in this introduction, this extension of thought experiments to mathematics was suggested by Ørsted, *Introduction*, 296–297, and Mach, *Knowledge*, 144–145. More recently, it was implied by Brown, *Laboratory*, 49–75. For the distinction between mathematics and physics in medieval thought experiments, see Grellard's contribution to this volume.

some concrete scenarios in mathematics that have appellations of the same kind—the Infinite Monkey Theorem, Buffon's needle problem, the Sleeping Beauty Paradox—but they do not represent mathematics as a whole, and it can be argued that they are only illustrations of abstract problems.

Thanks to these concrete scenarios, thought experiments have an attraction that we might call aesthetic and that may involve our imagination. They are, however, more like sketches than realistic paintings: we must strip them bare and isolate the significant details of the problem in question, so as to draw a conclusion that goes beyond the specific case described. For example, in the case of Galileo's two falling bodies strapped together, it is significant that the strapped bodies are not animated bodies (a baby and her mother, a cat and a dog), since animated bodies have spontaneous motions that modify their fall. But it is not significant that the bodies are cannonballs or musket balls, made of lead or made of wood, raw materials or polished works of art. There is nothing specific to thought experiments here: this is the case with real experiments as well, where, to give exaggerated examples, the star sign of the experimenter or the mileage on the car she has left in the lab's parking lot are not mentioned.

There is however a difference between experiments and thought experiments in one respect. In the reports of standardised experiments we find in today's science, all details mentioned are supposed to be significant with respect to the cognitive aim of the experiment. Thought experiments, on the other hand, include particulars that do not intervene in the outcome of the thought experiment itself, but which are mentioned in order to help us imagine the case. Hence, all the particulars involved in a thought experiment have not the same function: some of them are part of the argument itself, and some of them are just a way of making the argument psychologically convincing. For example, in one of the versions of Einstein's lift, the physicists are supposed to wake from narcotic sleep, during which they were presumably transported in the closed box where they observed bodies and exchanged theories.[62] One is tempted to say that this particular does not fundamentally change Einstein's thought experiment, but it helps us to imagine how it might be that two physicists can be found discussing gravitational fields in an opaque box.

[62] Einstein, "Present State of Gravitation," 208. The example is given by Norton, "Thought Experiments in Einstein," 130.

This seemingly elementary distinction between particulars that are part of the argument and particulars that are only a psychological crutch for our deficient imagination may be associated with a thesis on how thought experiments work. Indeed, the status of the particulars involved in the scenario may change whether one does or does not adhere to the inferentialist conception of thought experiments. In the first case, these particulars are irrelevant to the outcome of the thought experiment, since thought experiments are just arguments disguised in a pleasant way. In the second case, they are indispensable for the thought experiment to take place: it is due to the particulars that we may imagine ourselves in the situation depicted by the thought experiment and finally conclude what its outcome may be. Thus there is nothing surprising about Norton saying that the narcotic sleep of the physicists in the lift is irrelevant to the generality of the conclusion, while Nersessian judges that it reinforces crucial aspects of the thought experiment.[63]

Just as I asked, how far we want to go with weak counterfactuality, it is now time to note that, if we include any counterfactual concrete scenarios in the category of thought experiments, all sorts of fiction should be counted as thought experiments. At least philosophical and scientific fiction, like the depiction of the state of nature in early modern political philosophies, Descartes' genesis, or Plato's cave, but also our initial computer-generated scenario, the fabulous peregrinations of Lucian of Samosata, the transformation of a caliph into a beggar in *The Arabian Nights*, the voyage of Gulliver to the land of the Yahoos, the variable gender of the inhabitants of the planet Winter in *The Left Hand of Darkness*, or the ecological and political themes emerging in *Dune* ... These works of fiction indeed contain a good deal to think about, and, in certain cases, definitively have a speculative intention. Except if you beg the question by presupposing a difference between literary and non-literary thought experiments, you seem once again committed to include much more in thought experiments than you initially wished for. Hence the need for a final characteristic, namely that thought experiments are designed for framing a specific, preconceived thought goal, or that they have a well-determined cognitive intention.

---

[63] Ibid., 129; Nersessian, "In the Theoretician's Laboratory," 296.

2.3. In order to introduce this last characteristic, let us consider a well-known example coming from Bernard Williams' discussion with the Utilitarianist John Jamieson Carswell Smart.[64] Williams reproaches the fact that Carswell—and the Utilitarianists in general—set aside all questions arising from circumstances in which a moral agent can be found, and from the capacity of the moral agent to act effectively. He consequently depicts a situation of civil war, where a captain in a South American army makes Jim, a perfect stranger to the conflict, the offer to kill an Indian among a group of prisoners; if he does so, the other nineteen Indian prisoners will be freed; if he does not do it, they will all be shot. According to Williams, a genuine Utilitarianist will find it obvious that the stranger must accept the offer so as to save lives; but the way in which the Utilitarianist asks the question is problematic because it is disembodied; he does not put himself in the place of Jim. Through a vivid narrative, William invites his readers to identify themselves, at least temporarily, with Jim, and make his reactions their own; according to him, the readers who genuinely identify themselves with Jim will not easily adopt the Utilitarianist conclusion.

This thought experiment includes a counterfactual and concrete scenario that could very well figure in a novel written by Conrad. But obviously it is a thought experiment not only because of that, but also because Williams had a well-determined point concerning ethics. He wanted his readers to ask themselves not necessarily if it is possible to agree with the responses given by Utilitarianists, but if it is possible to be satisfied with the questions they ask. What confers the specificity to thought experiments when compared with any other fiction is that, just as crucial experiments are supposed to do, they clearly delimit a before and an after in what we know and what we think. Whether the point is about answering a question, raising a problem, testing a declaration by highlighting some paradoxes, or even giving proof of a previously unknown result, thought experiments have well-delimited cognitive intentions.[65] Thus, there is no absolute response to the question whether or not fiction should be included among thought experiments. It simply depends on the context in which it is inserted. Take, for example,

[64] Williams, "A Critique of Utilitarianism," 98–100. Williams' thought experiment is commented in Zeimbekis' essay.

[65] In her essay in this volume Ierodiakonou shows that ancient thought experiments start with an assumption that opens two possibilities to be explored.

Putnam's Brain-in-a-vat, or the Cartesian evil genius. These specific set-ups could be the beginning of literary fiction; insofar as they are inserted in argumentative contexts where well-delimited questions are asked, they are thought experiments.

To conclude this third characteristic, it should be noted that "having a well-determined cognitive intention" is related to "being an argument" without being identical with it. Just like thought experiments, arguments have well-delimited cognitive intentions, they clearly delimit a before and an after in our thinking, and they imply changes of beliefs. But it is a mistake to infer from the analogy between arguments and thought experiments to their identity. Similarly, it is a mistake to conclude from the terminal points of a process to the process itself. A belief can be expressed in a proposition; a change of beliefs can be seen as a transformation of one proposition into another; but the transformation itself does not necessarily result from propositions. Here again, it can be useful to think of real experiments. If I give up the belief that there is a butcher just around the corner for the belief that there is no butcher just around the corner, this change of beliefs may result from the fact that I have actually gone around the corner. This fact is not in itself propositional, even if, in explaining why I switch over to a new belief, I give the argument that I have actually gone around the corner. In other words, the well-determined cognitive intention of thought experiments is manifest in argumentative contexts, but to count the well-determined cognitive intention among the characteristics of thought experiments does not commit us to any position with regards to the inferentialist/non-inferentialist debate. On the contrary, it could easily be argued that thought experiments are necessary or simply useful because all that we know cannot be expressed as a chain of deductions.

I have explored the three characteristics of thought experiments. The benefit of this characterisation is twofold. Firstly, it easily succeeds in eliminating idealised experiments, mostly mathematical reasonings, and works of fiction not related to a well-determined cognitive intention. Of course, my point was not to eliminate for the sake of eliminating, but to make clear what criteria can either justify or invalidate the indefinite extension of the category of thought experiments that we have recently witnessed, and which undoubtedly has not yet come to an end. Secondly, this characterisation reveals that the category of thought experiments is affected perhaps not exactly by contradictions, but at least by tensions. For example, if a thought experiment has a well-delimited cognitive intention, why worry about concrete details that obviously do not change anything in its result? If a thought experiment is counterfactual, how can it allow us to know anything about the world?

3. *Nine Historical and Methodological Perspectives on Thought Experiments*

We can now understand why the category of thought experiment has been applied so wildly to different kinds of textual devices, and why some debates on thought experiments resembled a dialogue among deaf people. The crystallisation that Einstein and some other physicists of the 20th century brought about did not result in the constitution of a robust category; the protagonists of contemporary debates talked at cross purposes because they did not have the same characteristics of thought experiments in mind, or were not thinking of the same thought experiments. This does not mean that the category of thought experiment is not an interesting category: it is due to its vague boundaries that it raises questions.

This book was consequently guided by the idea that, since we cannot forge a strict definition of thought experiments, we should somehow tame the contemporary wild usage of this category by analysing the textual devices that we now retroactively qualify as thought experiments. All the essays in this volume have therefore both a historical dimension, since they rely on thought experiments borrowed from various periods in the history of philosophy, and a methodological dimension, in which they seek to clarify how thought experiments work, what their limits are, and what their conceptualisation could be.

The book is organised in three parts. The first one is devoted to the question of how to define and use the notion of thought experiments in different historical contexts, with essays by Ierodiakonou, Lautner, and Grellard. Howsoever thought experiments may be defined, different methodological objections can be raised regarding their very possibility: the second part deals with such objections, with essays by Knuuttila and Kukonnen, Palmerino, and Virvidakis. Finally, the essays by Engel, Goffi and Roux, and Zeimbekis try to provide a description of the way thought experiments are supposed to work.

3.1. As we have already explained, many questions have been asked in modern literature concerning thought experiments proposed in contemporary philosophy and in contemporary science. Are they distinct and is it possible to give a common definition? Are they heuristic devices, specific arguments, and paradoxes? Are they comparable to real experiments? But equally imaginative thought experiments are already found in ancient and medieval texts. The essays in the first part, provided by Ierodiakonou, Lautner, and Grellard, pay tribute to these prime historical examples of thought experiments, and show that historical perspectives help answering contemporary questions and, conversely, that these questions favour a better understanding of these texts.

Ierodiakonou focuses on the first recorded thought experiment, namely the thought experiment of the man who stands at the edge of the universe extending his hand or his stick. First, she explores the different versions of this thought experiment in antiquity: confronting Archytas' original version of this thought experiment with its Hellenistic appropriations, she examines its variations and shows that different philosophical schools took advantage of it. As it turns out, it was used to show that the world is indefinite (Archytas, according to Simplicius), it was related to the doctrine of the plurality of the worlds (Lucretius), or it was meant to prove that there is an infinite void beyond the heavens of our world (the Stoics, according to Posidonius). Furthermore, Ierodiakonou provides, based on Archytas' thought experiment, a tentative characterisation of ancient thought experiments. All of them imply a hypothetical scenario that starts with an imaginary assumption. If this assumption is the contrary of what the thought experiment is supposed to prove, then the thought experiment is indeed a *reductio ad absurdum* argument; but this is not invariably the case. Neither is it relevant whether the initial assumption is possible or not. The distinctive characteristic of ancient thought experiments, concludes Ierodiakonou, is that the assumption they start with opens two different possibilities that one needs to investigate, while it is not at all clear from the start which of the two one should follow.

Ierodiakonou looks for a formal characterisation of ancient thought experiments in general. The conclusion of Lautner's essay states that thought experiments might be differentiated according to the domains of knowledge in which they are introduced. Starting with a definition of thought experiments slightly different from the characterisation suggested in this introduction, he focuses on thought experiments presented in the *De anima* Neoplatonic commentaries written in the 6th century. According to the Platonic legacy, the soul was above the natural entities, but these commentaries admitted that it can be considered as a cause and that it can consequently be studied through its effects. In this context, effects are activities of the soul, such as sense-perception, and the strategy is to study these activities in order to understand the cause, that is, the essence of the soul. Lautner's point is that, contrary to what they did in natural philosophy, the commentators here drew on everyday experiences, completely free of counter-intuitive elements. Hence, the experiences that they resorted to in studying the nature of the soul can be qualified as thought experiments only if a very weak form of counterfactuality is allowed. After a detailed study of the commentators' treatment of the question, raised in *De anima* III 2, of what is responsible for the fact that we can perceive that we perceive, Lautner concludes that domain-specific considerations might be useful to sort out different kinds of ancient thought experiments. Weakly counterfactual thought experiments were indeed used in psychology and in ethics (e.g. Carneades' plank), while strongly counterfactual thought experiments were invented in physics (e.g. Archytas' stick) and in metaphysics (e.g. Theon and Dion).

Grellard's paper is devoted to what may be the golden age of thought experiments, namely medieval philosophy. As is well-known, medieval philosophers used their imagination to formulate hypotheses warranted by God's absolute power. In so far as they concern unobservable phenomena and as they imply a conceptual exploration, these hypotheses qualify as thought experiments. Grellard more precisely focuses on thought experiments that Henry of Harclay formulated in the 14th century in order to establish the existence of atoms, i.e. the existence of entities without extensions, which are the ultimate constituents of a body. The first thought experiment involves a sphere moving continuously on a plane, the second one relies on God's ability to see all the points in a line. Grellard then examines the arguments that Adam Wodeham, William of Alnwick, John Buridan, and Nicholas Oresme, who did not believe in the existence of atoms, opposed to these two thought experiments. Their arguments were twofold. On the one hand, in order to refute these thought experiments, they exhibited their logical flaws and questioned their physical plausibility. On the other hand, in order to limit their scope, they argued that they should be considered as abstract mathematical reasonings, which do not pertain to physical realities. What is revealed by the practice of thought experimenting in medieval debates on atomism, Grellard concludes, is the problem of articulating natural philosophy and mathematics.

3.2. Whatever thought experiments are considered, objections can be raised about their very possibility. Since all thought experiments begin with counterfactual suppositions, and sometimes with impossible hypotheses, the first question is what kind of conceptualisation can account for such a process. We should also wonder if counterfactual considerations can teach us anything about our actual real world. Lastly, the question of tracing the borders between legitimate and illegitimate thought experiments and of explaining the plausibility of certain thought experiments is raised. In the second part of the book, the essays by Knuuttila and Kukkonen, Palmerino and Virvidakis address those three methodological questions from a historical perspective.

Knuuttila and Kukkonen's essay deals with an important question for the methodology of thought experiments, namely the question of the kind of conceptualisation that can be given of thought experiments starting with impossible premises. Their starting point is the fact that, in Aristotle's reductive proofs, there are many arguments with premises that involve hypotheses in contradiction with principles elsewhere described as necessary. However, Aristotle never fully explained the status of these impossible premises, except when noting that impossibilities can be assumed to be possible in some sense. The problem has already been discussed in antiquity, and likewise medieval commentators of the 12th and 13th centuries tried to find a solution. Averroes argued that the impossible premises are impossible *per accident*, but possible essentially or *per se*: they are false in our world, but the nature of things does not demand their falsity. Averroes' abstractionist approach of thought experiments was later expanded by Aquinas, who insisted that something which is possible for a thing as a member of a genus can be impossible for it as a member of a species, the same holding for a member of a species and an individuated being. Hence, Knuuttila and Kukkonen continue, Averroes and Aquinas developed an earlier conceptualisation of thought experiments alternative to the one that is famously illustrated by Buridan and Oresme. Namely, it is usually admitted that the conceptual background of medieval thought experiments was the new modal theory developed in the early 14th century, the reference to God's omnipotence, and the notion that possible worlds could include features quite divergent from ours. However, neither Averroes nor Aquinas relied on this conceptual background: Knuuttila and Kukkonen's conclusion is that their conceptualisation might have been closer to Aristotle's way of thinking.

Palmerino shows that many thought experiments presented in Galileo's works derived from medieval and Renaissance sources. However, she argues, this must not be read as a sign of continuity between medieval philosophy and early modern science: rather, Galileo used a traditional way of arguing in order to build up non-traditional conclusions. More precisely, she divides the medieval thought experiments that reappear in Galileo's works into two groups: those that concerned the actual physical world, and those that hinged on worlds which God, in his infinite potency, could have created. According to her, arguments of the first type (e.g. Buridan's arrow, or Albert of Saxony's stone, which falls through a tunnel passing by the centre of the earth), served Galileo to show that medieval natural philosophers had detected some major inconsistencies in Aristotle's physics, but had not understood the disruptive implications of their critique. Arguments of the second type (concerning for example the centre of the universe or the possibility of a void) were instead used to show that what medieval authors had imagined to happen in alternative worlds could actually be the case in our own world. But, Palmerino concludes, Galileo explicitly rejected thought experiments that appealed to divine omnipotence; according to him, nothing meaningful can be said about a situation, in which the laws of physics are violated and the only way to reason about imaginary scenarios is to explain them in conformity with the ordinary course of nature.

Just like Knuuttila and Kukkonen, Virvidakis makes use of the history of philosophy to understand the methodology of thought experiments. Indeed, he takes on the challenge formulated by Brown to provide us with a fully-fledged Kantian account of thought experiments that would constitute an alternative to the existing empiricist and Platonic accounts of thought experiments.[66] However, Virvidakis' discussion aims not so much at providing this alternative than rather suggesting that Kant's transcendental perspective indicates general criteria for the evaluation of thought experiments. Virvidakis first examines Kant's objections to thought experiments proposed by pre-critical thinkers: the problem with Descartes, Leibniz and Hume is [in his opinion] that they ignore the epistemic conditions of our mental faculties; in particular, they are not able to distinguish conceivability or imaginability and real possibility. Furthermore, Virvidakis shows that Kant's transcendental enterprise is guided by questions about the possibility of experience, and by the analysis of modalities. As a conclusion, he extends the Kantian lesson to all thought experiments, regardless of the area and topics dealt with: in order to know which thought experiments we should trust, we should distinguish between logical possibility concerning *entia rationis*, and real

[66] Brown, *Laboratory*, 156.

or transcendental possibility, which is essential for reaching conclusions about the world as it appears to us.

3.3. Finally, the last section tries to provide a description of the way thought experiments work. Of course, such a description of the functioning of thought experiments depends on where we draw the lines between thought experiments and other cognitive procedures, and this in turn may depend on the kind of thought experiments that are considered. The essays by Engel, Goffi and Roux, and finally Zeimbekis do not actually focus on the same aspects of thought experiments, or, once again, on the same thought experiments; Engel focuses on the modal claims involved in thought experiments; Goffi and Roux consider thought experiments in so far as they are arguments; Zeimbekis investigates the viability of a simulationist account of thought experiments.

Drawing on classical examples of thought experiments (Lucretius' spear and Jackson's Mary), Engel suggests that every thought experiment involves a modal claim concerning the possibility of the situation described in the thought experiment and an epistemological claim concerning the knowledge we have of this situation. This leads him to review the existing epistemological conceptions of thought experiments: according to him, the debate cannot be settled unless the modal claims involved in them are clarified. It is at this point that he introduces the central thesis, recently advocated by Timothy Williamson about the Gettier cases, according to which the modal claims involved in philosophical thought experiments are just ordinary counterfactual reasonings. Engel agrees with Williamson's thesis and sees it as a way of eliminating two symmetrical opposite views about thought experiments: the ultra-rationalist view, according to which they appeal to *a priori* intuitions, and the ultra-empiricist view, according to which they appeal to empirical beliefs only. He also agrees with Williamson that Gettier's cases bear upon knowledge, not upon our concept of knowledge. However, contrary to Williamson, he argues that the existence of conflicting conceptions about knowledge entails that thought experiments are partly conceptual and partly metaphysical.

Like Virvidakis, but from an entirely different perspective, since their approach does not require the examination of transcendental conditions of experience, Goffi and Roux are interested in what makes some thought experiments work, while others do not work. They do not attempt to draw an *a priori* line between two types of thought experiments, but rather ask the following question: inasmuch as thought experiments are arguments, and notwithstanding the fact that some of them might involve the contemplation of an imaginary scenario, how is it that some of them work, while others do not? Taking inspiration from a counterfactual thought experiment presented by Nicholas Rescher, they treat thought experiments as argumentative procedures resembling tests of consistency, which invite the experimenter to seek the weakest link in her body of beliefs. Equipped with this method, they examine two well-known successful thought experiments (Galileo's two bodies strapped together, and Thomson's violinist) and discuss Mach's notion of thought experiments. Thus they reach the hypothesis that successful thought experiments respect the three following conditions: they do not deal with things, but with beliefs; they mobilise a set of beliefs shared by the interlocutors; and this set of beliefs has a hierarchical structure. Using once again examples written at different periods and taken from various disciplines (Descartes' receding bodies, Aristotle's weaving shuttles), Goffi and Roux argue that each of those conditions is individually necessary for a thought experiment to work. They finally conclude on the limits and consequences of their approach.

Finally, Zeimbekis tests a number of different simulationist conceptions of thought experiments, presented by Gendler, Nersessian, and Gaut. According to those accounts, in thought experiments the mind somehow simulates processes with which it reaches conclusions. Zeimbekis's argument relies on a clear-cut distinction between thought experiments dealing with physical situations, and thought experiments dealing with mental situations. In the first case, he argues, mostly against Gendler, that thought experiments cannot simulate physical processes, but rather use mental models. In the second case, he focuses on moral thought experiments; commenting on Williams' "Jim and the Indians"-thought experiment, he shows that Williams, when he described this concrete scenario for us to simulate it mentally, *de facto* favoured moral theories that are committed to an egocentric viewpoint. Hence, the scope of mental simulation in thought experiments is primarily limited by the constraint of relevant similarity on source and target processes: on one hand, this constraint disqualifies thought from simulating external natural processes; on the other hand, it is a source of epistemic bias in moral thought experiments. Zeimbekis consequently concludes that thought experiments and mental simulations cannot be assimilated as means of acquiring knowledge.

PART ONE

HISTORICAL USES OF THOUGHT EXPERIMENTS

REMARKS ON THE HISTORY OF AN ANCIENT THOUGHT EXPERIMENT

Katerina Ierodiakonou

Philosophers tend to be rather conservative when it comes to choosing examples to illustrate and back up their theories. That is to say, there are many examples in the history of philosophy, which have been used again and again by different philosophers in different periods, sometimes for the same and sometimes for different purposes. For instance, the example of the square large tower which looks small and round at a distance is mentioned by the Epicureans and the Stoics as well as by Descartes. Similarly, the example of the man who suffers from jaundice and perceives everything as yellow is found both in Sextus Empiricus' and in Hume's writings, though one would expect that philosophers at some early point could have recognised its implausibility and abandoned it. Indeed, this conservative attitude does not only apply to the case of examples. It is even more intriguing that philosophers decide to also borrow thought experiments from previous philosophers, often from philosophers who belong to a different philosophical tradition with a very different outlook. And I think that the case of the re-used thought experiments is even more intriguing than that of the borrowed examples, because a thought experiment does not primarily depend on our common experience; rather, it is the product of the imagination of a specific philosopher who constructs it sometimes in order to prove, and not simply to illustrate, a certain philosophical theory.

There are, in fact, some thought experiments which were obviously perceived as very illuminating by philosophers with very different positions, since they keep on reappearing in different periods in the history of philosophy. The first recorded thought experiment, the ancient thought experiment of the man who stands at the edge of the universe extending his hand or his stick, certainly is such a case. It was initially meant to prove the infinity of the universe and it is first attributed to the Pythagorean Archytas of Tarentum in the first half of the 4th century; it was then appropriated by the Epicureans and the Stoics during the Hellenistic period, while in late antiquity the Aristotelian commentators discussed its uses in detail; in the Middle Ages, Thomas Aquinas, followed by a number of 14th century scholars, such as John Buridan, Nicholas Oresme and Richard of Middleton, depended on its Stoic version and at the same time elaborated on it in the light of their Christian beliefs; finally, in early modern times, Pierre Gassendi, Otto von Guericke, John Locke and Isaac Newton also referred to it, each for his own reasons.[1]

In my paper I want to focus, in particular, on how the Hellenistic philosophers used this thought experiment, and to address the following two questions:

1. Does the original version of this thought experiment significantly differ from its Hellenistic appropriations?
2. Can we, on the basis of this thought experiment, get clearer about the general characteristics of thought experiments as used by philosophers in antiquity?

Let me start with the surviving evidence about Archytas' thought experiment, which is to be found in Simplicius' commentary on the *Physics*, at the end of the discussion of Aristotle's five arguments against the existence of something infinite (*in Phys.* 467.26–35; trans. C.A. Huffman):[2]

> "But Archytas," as Eudemus says, "used to propound the argument in this way: 'If I arrived at the outermost edge of the heaven [that is to say at the fixed heaven], could I extend my hand or staff into what is outside or not?' It would be paradoxical not to be able to extend it. But if I extend it, what is outside will be either body or place. It doesn't matter which, as we will learn. So then he will always go forward in the same fashion to the limit that is supposed in each case and will ask the same question, and if there will always be something else to which his staff [extends], it is clear that it is also unlimited. And if it is body, what was proposed has been demonstrated. If it is place, place is that in which body is or could be, but what is potential must be regarded as really existing in the case of eternal things, and thus there would be unlimited body and space."

---

[1] See Grant, "Medieval and Seventeenth-century Conceptions"; idem, *Much Ado about Nothing*.

[2] "'Ἀρχύτας δέ", ὥς φησιν Εὔδημος, "οὕτως ἠρώτα τὸν λόγον· 'ἐν τῷ ἐσχάτῳ οἷον τῷ ἀπλανεῖ οὐρανῷ γενόμενος, πότερον ἐκτείναιμι ἂν τὴν χεῖρα ἢ τὴν ῥάβδον εἰς τὸ ἔξω, ἢ οὔ;' καὶ τὸ μὲν οὖν μὴ ἐκτείνειν ἄτοπον· εἰ δὲ ἐκτείνω, ἤτοι σῶμα ἢ τόπος τὸ ἐκτὸς ἔσται. διοίσει δὲ οὐδὲν ὡς μαθησόμεθα. ἀεὶ οὖν βαδιεῖται τὸν αὐτὸν τρόπον ἐπὶ τὸ ἀεὶ λαμβανόμενον πέρας, καὶ ταὐτὸν ἐρωτήσει, καὶ εἰ ἀεὶ ἕτερον ἔσται ἐφ' ὃ ἡ ῥάβδος, δῆλον ὅτι καὶ ἄπειρον. καὶ εἰ μὲν σῶμα, δέδεικται τὸ προκείμενον· εἰ δὲ τόπος, ἔστι δὲ τόπος τὸ ἐν ᾧ σῶμά ἐστιν ἢ δύναιτ' ἂν εἶναι, τὸ δὲ δυνάμει ὡς ὂν χρὴ τιθέναι ἐπὶ τῶν ἀιδίων, καὶ οὕτως ἂν εἴη σῶμα ἄπειρον καὶ τόπος."

Modern scholars generally agree that this thought experiment can be rightly attributed to Archytas. There are some differences in opinion, though, as to how exactly Archytas originally formulated it, what precisely he wanted to prove, and whose theory he was trying to refute.[3] It has been suggested, for instance, that there are elements in Simplicius' report that indicate that Archytas' original thought experiment was recast by Eudemus so that it better fitted his own Aristotelian concerns. Also, it has proved more complicated to determine whether Archytas was interested in demonstrating that the universe is infinite or simply indeterminate. Finally, it is still controversial whether this thought experiment was meant as an argument against Parmenides' finite sphere or against the finite cosmos of Plato's *Timaeus*.

Important as they may be, in what follows I do not want to investigate these issues concerning Archytas' intentions in suggesting this experiment. Rather, I want to outline the general structure of Archytas' thought experiment and keep it as the point of reference for my subsequent account of its later versions: imagine, Archytas said, that the universe is finite, that there is a man at its outermost edge, and that he tries to extend his hand or his stick. There are two possibilities: either he will or he won't be able to extend it. But it is absurd to think that he won't be able to extend it. And if he will be able to extend it, it means that he is not at the edge of the universe; the same question can thus be raised when he moves further out, and this question can continue to be raised *ad infinitum*. Hence, if we assume that the universe is finite, we either reach an absurdity or conclude that the universe is infinite. Therefore, the universe is infinite.

There is no evidence that Archytas' thought experiment was used by other philosophers up until the Hellenistic period, and this should not be surprising. The two main philosophical traditions of the classical period for which we have ample evidence, namely Platonism and Aristotelianism, rejected the view that the universe is infinite; it is thus reasonable to think that they had no use for a thought experiment designed to support such a claim. On the other hand, when Epicurus and the Epicureans again began to try to defend the thesis that the universe is infinite, Archytas' thought experiment re-emerged. More specifically, in the 1st century BC, Lucretius, who on the whole or altogether relied on Epicurus' *On nature*,

[3] See Sorabji, *Time, Creation and the Continuum*, 125–128; Huffman, *Archytas of Tarentum*, 540–550.

used the same thought experiment in his philosophical poem *De rerum natura* (1.968–983; trans. M.F. Smith):[4]

> Then again, just suppose that all the existing space were finite, and that someone ran forward to the edge of its farthest border and launched a spear into flight: do you favour the view that the spear, cast with virile vigour, would fly far and reach its target, or do you suppose that something could check it by obstructing its course? You must grant and adopt one or the other of these hypotheses, and yet both deny you a subterfuge and compel you to acknowledge that the expanse of the universe is infinite. For whether there is something to check the spear and prevent it from hitting its mark and lodging in its target, or whether it flies on, it did not start from the end of the universe. In this way I will dog you: whenever you locate the farthest border, I will ask about the ultimate fate of the spear. Our conclusion will be that nowhere can a boundary be fixed: no escape will ever be found from the limitless possibility of flight.

This text comes right after an argument in favour of the view that the universe is infinite; an argument also found in Epicurus' own extant writings. Lucretius (*De rerum natura* 1.958–967), following Epicurus (cf. *Ep. ad Hdt.* 41), claims that the universe is infinite, because otherwise it would have to have an end; but nothing has an end, unless there is something beyond it to bind it, and since there is nothing outside the universe, the universe has no end. The thought experiment of the man at the edge of the universe who throws his spear is presented as the second argument in favour of the thesis that the universe is infinite, though it could be seen as an illustration of the first argument. Moreover, it is not entirely clear whether Lucretius found this thought experiment in an Epicurean text or whether he added it himself. It has been pointed out

[4] Praeterea si iam finitum constituatur
omne quod est spatium, siquis procurrat ad oras
ultimus extremas iaciatque volatile telum,
id validis utrum contortum viribus ire
quo fuerit missum mavis longeque volare,
an prohibere aliquid censes obstareque posse?
alterutrum fatearis enim sumasque necessest;
quorum utrumque tibi effugium praecludit et omne
cogit ut exempta concedas fine patere.
nam sive est aliquid quod probeat officiatque
quominu' quo missum est veniat finique locet se,
sive foras fertur, non est a fine profectum.
hoc pacto sequar atque, oras ubicumque locaris
extremas, quaeram quid telo denique fiat.
fiet uti nusquam possit consistere finis
effugiumque fugae prolatet copia semper.

that the military terms used here by Lucretius give it a Roman touch and suggest that this version of Archytas' thought experiment could well have been his own invention. In particular, the use of the term "*procurrat*," meaning the run before the throwing of a spear, and of the term "*contortum*," referring to the actual throwing of a spear, point to a Roman military context as found, for instance, in Virgil's *Aeneid* (cf. 2.50; 12.267). Besides, it is worth keeping in mind that the hurling of a spear is associated with an old Roman custom, according to which a certain priest would go to the boundaries of the town and hurl a spear as a sign for the declaration of war.[5] Yet it could also be Lucretius' Roman version of an Epicurean argument that talked about stretching out a hand or a stick or a spear.

But whether this thought experiment is by Lucretius himself or belongs to the Epicurean tradition well before him, what is important for our purposes is to examine what differentiates this version from Archytas' original thought experiment. So, let me sketch Lucretius' thought experiment: imagine that the universe is finite, that there is a man at its outermost edge, and that he tries to throw a spear. There are two possibilities: either the spear will follow its natural trajectory unhindered, or it will be stopped by a body and bounce back. But in either case the man throwing the spear was not at the edge of the universe; the same question can thus be raised when the man moves further out, and this question can be raised *ad infinitum*. Hence, if we assume that the universe is finite, the thought experiment leads us to the conclusion that the universe is infinite. Therefore, the universe is infinite.

There is no doubt that the most conspicuous diversion of the Epicurean thought experiment in Lucretius' version from Archytas' original version is that the man throws a spear instead of extending his hand or his stick. This obvious difference can of course be attributed to Lucretius' rhetorical skills; for it is reasonable to suggest that he chose to modify Archytas' thought experiment in order to make it more vivid and picturesque for a Roman audience. Yet, there might be something more to account for this difference. That is to say, by assuming that the man throws a spear instead of extending his hand or his stick, the thought experiment does not lead to a gradual expansion of the universe; the spear aims at an unspecified target, which could be close or far away,

[5] See Bailey, *De rerum natura*, 766–767; Bignone, *Storia della Letteratura Latina*, 229–230.

it could be in our world or in another world. Thus it seems that Lucretius' thought experiment is purposefully designed in such a way that it not only supports the view that our world does not have a fixed limit, but it also is sensitive to the Epicurean doctrine that there are infinitely many worlds.

However, neither Archytas' original thought experiment nor its Epicurean modification seem to have been the most influential in the history of philosophy; it is mainly the Stoic version of this thought experiment that most of the medieval and early modern philosophers refer to in their writings. And this is because the Stoic version was preserved in Simplicius' commentary on Aristotle's *De caelo*, a commentary translated into Latin by William of Moerbeke in 1271,[6] whereas Simplicius' commentary on the *Physics* was not translated into Latin until the 16th century (*in De cael.* 284.28–285.2; trans. R.J. Hankinson):[7]

> The Stoics, however, wanted there to be void outside the heaven, and they established this through the following supposition. Suppose, they say, someone stands at the extremity of the sphere of the fixed [stars] and extends his hand upwards. If he does extend it, they infer that there is something outside the heaven into which he extends it; but if he cannot extend it, there will be something outside in such a way as to prevent his hand's extension. And if he then stands at the limit of this and extends, the same question [recurs]. For something will be shown to exist outside this as well.

So, the Stoic version runs as follows: imagine that the universe is finite, that there is a man at its outermost edge, and that he tries to extend his hand. There are two possibilities: either he will be able to extend it, which means that there is something beyond the heavens, or he won't be able to extend it, which means that there is something obstructing it; the same question can thus be posited when the man moves to the obstacle. Therefore, there is something beyond the heavens.

The Stoics used the same thought experiment that Archytas had used long before them, but just like Lucretius they chose to further elaborate

---

[6] See Grabmann, *Guglielmo di Moerbeke*.

[7] Οἱ δὲ ἀπὸ τῆς Στοᾶς ἔξω τοῦ οὐρανοῦ κενὸν εἶναι βουλόμενοι διὰ τοιαύτης αὐτὸ κατασκευάζουσιν ὑποθέσεως. ἔστω, φασίν, ἐν τῷ ἐσχάτῳ τῆς ἀπλανοῦς ἑστῶτά τινα ἐκτείνειν πρὸς τὸ ἄνω τὴν χεῖρα· καὶ εἰ μὲν ἐκτείνει, λαμβάνουσιν, ὅτι ἔστι τι ἐκτὸς τοῦ οὐρανοῦ, εἰς ὃ ἐξέτεινεν, εἰ δὲ μὴ δύναιτο ἐκτεῖναι, ἔσται τι καὶ οὕτως ἐκτὸς τὸ κωλῦσαν τὴν τῆς χειρὸς ἔκτασιν. κἂν πρὸς τῷ πέρατι πάλιν ἐκείνου στὰς ἐκτείνῃ, ὁμοία ἡ ἐρώτησις· εἶναι γὰρ δειχθήσεται κἀκείνου τι ἐκτὸς ὄν.

on one of the two possibilities, namely the possibility according to which the man won't be able to extend his hand; for, instead of stating that it is absurd to claim that he won't be able to extend it, they suggested that in this case there must be something which obstructs it. More importantly, the Stoics seem to have used the same thought experiment for a different purpose. They were not interested in proving that the universe is infinite, but that there is infinite three-dimensional void beyond the heavens surrounding our finite world. For, according to Stoic cosmology, the existence of such a void is a necessary prerequisite for the possibility of the expansion of the world in the course of each conflagration, when the entire corporeal substance of the world is resolved utterly into fire and, fire becoming purer and purer and hence more and more rarefied, expands in volume (e.g. Cleomedes I.1.43–54 Todd). However, such a cosmological theory leaves it open, whether the void beyond the heavens should be taken to be infinite or simply indefinite in extension.

There are ancient sources attributing to Posidonius (F97a, b Edelstein and Kidd) the position that the void outside the world is not infinite but sufficiently large for the conflagration. On the other hand, there are reliable texts providing us with Stoic arguments in favour of the infinity of void; for instance, Cleomedes (I.1.106–109 Todd) argues that if the void is not infinite, there should be a body to limit it, but there could be no body outside the void limiting it, and thus the void should be infinite. Modern scholars have tried to resolve this inconsistency in different ways.[8] It is plausible to suggest, I think, that the issue here concerns the different senses of the Greek term "*apeiron*," which may be understood as meaning "infinite" but also as meaning "indefinite." Thus, although Posidonius seems to have insisted that the void outside the world is *apeiron* only in the sense of being indefinite, other Stoics were obviously of a different opinion. In this case the Stoics' use of Archytas' thought experiment may have proved confusing, if not misleading; for it may have played an important role in attributing to the Stoic school as a whole the view that the void outside the world is infinite rather than indefinite, since this particular thought experiment had previously been used both by Archytas and by the Epicureans to prove the infinity of the universe. Is this the price to pay when one borrows thought experiments from past philosophers belonging to different schools?

[8] See Algra, *Concepts of Space in Greek Thought*; Hahm, *The Origins of Stoic Cosmology*.

But let us return again to Archytas' thought experiment and closely study its structure in virtue of which it is classified, at least according to our modern understanding, as a thought experiment.

The first striking feature of Archytas' thought experiment is, of course, its hypothetical scenario: imagine that the universe is finite, that a man stands at its outermost edge, and that he extends his hand or his stick. This hypothetical scenario assumes the contradiction of the thesis supposed to be proven by the thought experiment, namely that the universe is infinite; in other words, this thought experiment has the structure of a *reductio ad absurdum* argument. But were all thought experiments in antiquity *reductio ad absurdum* arguments? Certainly not. For instance, to focus again on the Hellenistic period, there is another famous thought experiment used by the Stoics that does not have such a structure. According to Philo's *De aeternitate mundi* (48), Chrysippus asks us to imagine two individuals, namely Dion and Theon; Dion is a whole-limbed human being, while Theon is part of Dion in the sense that he is defined in all aspects just like Dion, except that he lacks a leg. Then, we are supposed to consider what happens if Dion loses the leg that Theon is lacking. Is it that Dion perishes and Theon survives, as Philo claims, or is it that Dion survives and Theon ceases to exist, as Chrysippus tried to prove? There are many difficulties in interpreting this thought experiment; there is no doubt, however, that it is not cast as a *reductio ad absurdum* argument.

Another conspicuous characteristic of Archytas' thought experiment is that its hypothetical scenario starts with an impossible assumption; for no ancient philosopher seems to have thought that it is actually possible for a man to stand at the outermost edge of the universe. It is worth noting, though, that much later in the history of philosophy, for instance in John Locke's version, the initial assumption of this thought experiment was presented as possible; for it was said that it is God himself, who places the man of our thought experiment at the outermost edge of the universe (cf. *Essay*, Book II, ch. 13 § 21). Still, the fact that Archytas' thought experiment starts with an impossible assumption does not mean that all thought experiments in antiquity had to start in this way. On the contrary, it seems that there were ancient philosophers, who argued that it is not permissible to use impossible assumptions in order to construct thought experiments that are meant to prove a philosophical claim regarding the world. The text in Simplicius' commentary, which comes immediately after his presentation of the Stoic version of Archytas' thought experiment, is very interesting in this regard. For according to

Simplicius' report, Alexander of Aphrodisias strongly objected to the use of this particular thought experiment partly because it was based on an impossible assumption.

More specifically, Simplicius claims that Alexander had a convincing argument against the use of the Stoic version of Archytas' thought experiment. According to Alexander's argument, what prevents the man's hand from being extended is not some body at the furthest edge of the world to which the man could move and then give rise again to the same question; rather, there is nothing outside our finite world, and it is absurd to think that the man will be able to extend his hand into nothing (*in De cael.* 285.21–27; cf. Alexander, *Quaest.* 3.12, 106.32–107.4). But as I have already explained prior to presenting this argument, Simplicius also refers to another objection that Alexander raised against the Stoic version of Archytas' thought experiment, namely that its initial assumption is impossible (*in De cael.* 285.2–9; trans. R.J. Hankinson):[9]

> 'That this is unsound,' Alexander says, 'may be shown thus. If the world is the universe, and there is nothing outside the universe, there will be nothing outside the world. For it would not be the universe if there were something else outside it. Consequently no one would be able to extend their hand having arrived at the extremity of the heaven. Further, it is not at all possible and contrary to supposition for something to come to be there. For the divine body that moves in a circle is unaffectible and incapable of admitting anything of such a kind within itself.'

And some paragraphs later (*in De cael.* 286.23–27; trans. R.J. Hankinson):[10]

> Alexander also attacks those who consider what has been said to be true, following the imaginary case of the hand. 'For,' he says, 'we imagine many impossible things; and each person may imagine himself outside the town, and many times his size, or much smaller and the size of a millet-seed.'

Simplicius, however, disagrees with Alexander on this matter and defends the practice of using hypothetical or "imaginary," as he himself

[9] "ὅτι δέ, φησὶν Ἀλέξανδρος, μὴ ὑγιὲς τοῦτο, οὕτως ἂν δειχθείη· εἰ ἔστιν ὁ κόσμος τὸ πᾶν, καὶ μηδὲν ἔστι τοῦ παντὸς ἐκτός, οὐδ' ἂν τοῦ κόσμου τι ἐκτὸς εἴη· οὐ γὰρ ἂν ἔτι πᾶν εἴη, εἴ τι εἴη ἐκτὸς ἄλλο αὐτοῦ· ὥστε οὐδ' ἂν δύναιτό τις ἐκτεῖναι τὴν χεῖρα γενόμενος ἐπὶ τῷ ἐσχάτῳ τοῦ οὐρανοῦ· καὶ γὰρ τὸ ὅλως ἐκεῖ γενέσθαι τινὰ ἀδύνατον καὶ πρὸς ὑπόθεσιν· τὸ γὰρ σῶμα τὸ κύκλῳ κινούμενον τὸ θεῖον ἀπαθὲς καὶ μηδὲν τοιοῦτον ἐν αὑτῷ δέξασθαι δυνάμενον."

[10] διαβάλλει δὲ ὁ Ἀλέξανδρος καὶ τὸ τῇ φαντασίᾳ τῆς χειρὸς ἀκολουθοῦντας ἀληθὲς ἡγεῖσθαι τὸ λεγόμενον. "πολλῶν γάρ, φησί, καὶ ἀδυνάτων φαντασίας ἴσχομεν· καὶ γὰρ ἕκαστος ἑαυτὸν ἔξω τοῦ ἄστεος φαντασθείη ἂν καὶ τὸ μέγεθος πολλαπλάσιον ἢ βραχύτατον καὶ κεγχριαῖον."

calls them, though impossible, scenarios in order to construct thought experiments. Indeed, he invokes for this purpose the authority of Plato and Aristotle, both of whom are said to have used thought experiments with imaginary scenarios in the *Timaeus* (63B–C) and in the *De caelo* (310b3–4) respectively (*in De cael.* 285.9–21; trans. R.J. Hankinson):[11]

> But this objection is perhaps not conclusive, since we often suppose impossibilities for the sake of the subsequent consequence of the argument. Plato indeed supposed someone to be standing on the fire-sphere where the fire exists in its purest nature, having his head inclined towards the earth, and having the ability to abstract some of the fire and place it in the scales and drag the balance towards the air. And although this is impossible, he supposed it in order to show that down and up, heavy and light, do not exist by nature at the sphere. Indeed he concludes from the supposition that 'when two things are raised at the same time by a single force, the smaller necessarily yields more and the larger less to what constrains them by force, and the former is called heavy and downward-moving, and the smaller light and upward-moving. And Aristotle in the fourth book will suppose the earth to be transferred to the place of fire.

Interestingly enough, this is a view which we also find in Philoponus' commentary on Aristotle's *Physics*,[12] though the examples taken here from Plato's *Timaeus* (53B) and Aristotle's *De caelo* (295a21–29) differ (*in Phys.* 575.2–11; trans. D. Furley):[13]

---

[11] ἀλλ' αὕτη μὲν ἴσως οὐκ ἀναγκαία ἡ ἔνστασις· πολλάκις γὰρ ὑποτιθέμεθα τὰ ἀδύνατα διὰ τὴν ἐφεξῆς ἀκολουθίαν τοῦ λόγου· καὶ ὁ Πλάτων γοῦν ὑπέθετό τινα ἐν τῷ ὑπεκκαύματι ἑστῶτα, καθ' ὃ ἡ τοῦ πυρὸς μάλιστα φύσις, καὶ πρὸς τῇ γῇ νεύουσαν ἔχοντα τὴν κεφαλὴν καὶ δύναμιν λαβόντα ὥστε τοῦ πυρὸς ἀφαιρούμενον ἱστάναι εἰς πλάστιγγας καὶ ἕλκειν τὸν ζυγὸν ἐπὶ τὸν ἀέρα. καίτοι καὶ τοῦτο ἀδύνατον, ἀλλ' ἵνα δείξῃ μὴ φύσει ὂν τὸ κάτω καὶ τὸ ἄνω καὶ τὸ βαρὺ καὶ τὸ κοῦφον ἐπὶ τῆς σφαίρας, τοῦτο ὑπέθετο. ἐπάγει γοῦν τῇ ὑποθέσει· "ῥώμῃ γὰρ μιᾷ δυοῖν ἅμα μετεωριζομένοιν τὸ μὲν ἔλαττον μᾶλλον, τὸ δὲ πλέον ἧττον ἀνάγκη που κατατεινόμενον ξυνέπεσθαι τῇ βίᾳ, καὶ τὸ μὲν βαρὺ καὶ κάτω φερόμενον κληθῆναι, τὸ δὲ σμικρὸν ἐλαφρὸν καὶ ἄνω". καὶ ὁ Ἀριστοτέλης δὲ ἐν τῷ τετάρτῳ βιβλίῳ μετατιθεμένην τὴν γῆν εἰς τὸν τοῦ πυρὸς τόπον ὑποθήσεται.

[12] See Martin, "Non-reductive Arguments from Impossible Hypotheses."

[13] οὐ γὰρ ἄτοπον ὑποθέσεσι χρῆσθαί τισιν, εἰ καὶ ἀδύνατοι εἶεν, πρὸς ἄλλων ἐπίσκεψιν· ἐντεῦθεν γὰρ ἐπειδὴ τῇ κινήσει τοῦ οὐρανοῦ τινες ἀπεδίδοσαν τὴν αἰτίαν τοῦ ἵστασθαι τὴν γῆν, διὰ τοῦτό φησιν ὁ Ἀριστοτέλης, ὅτι ἵνα ἴδωμεν εἰ τῷ ὄντι ἡ τοῦ οὐρανοῦ κίνησις αἰτία ἐστὶ τοῦ ἵστασθαι τὴν γῆν, στήσωμεν τῷ λόγῳ τὸν οὐρανόν, καὶ ζητήσωμεν ποῦ οἰσθήσεται κατὰ φύσιν ἡ γῆ. καὶ ὁ Πλάτων δὲ τὸν τῆς τάξεως τοῦ παντὸς αἴτιον κατ' ἐπίνοιαν τοῦ κόσμου χωρίσας, ζητεῖ πῶς ἂν ἔχοι τὸ πᾶν αὐτὸ καθ' αὑτὸ θεοῦ χωρισθέν. καίτοι τούτων ἕκαστον ἀδύνατον γενέσθαι, ἀλλ' ὅμως τὰ τῇ φύσει συνημμένα χωρίζει κατ' ἐπίνοιαν ὁ λόγος, ὡς ἂν ἴδοι πῶς ἕκαστον αὐτὸ καθ' αὑτὸ κατὰ τὴν ἰδίαν ἔχει φύσιν.

> It is not absurd to use hypotheses even if they are impossible, for the examination of other things; on this ground, since some people explained the stationary position of the earth by the motion of the heaven, Aristotle says that we should stop the heaven in theory, and ask where the earth will go by nature, so as to see whether the movement of the heaven really is the cause of the stationary position of the earth. Plato, too, separating in thought the cause of the order of the universe from the cosmos, asks how the universe would be, all by itself and without God. Each of these cannot possibly come about; all the same, theory separates in thought things naturally conjoined, so as to see how each thing is by itself in its own nature.

But even if we disregard Alexander's objections, an impossible hypothetical scenario cannot be said to be a necessary feature of an ancient thought experiment. Let me mention another example of a thought experiment, which was also used during the Hellenistic period and does not depend on an impossible assumption. To prove the Epicurean thesis that death is nothing to us, Lucretius (*De rerum natura* 3.847–861) asks us to imagine that all our atoms are reassembled after our death; even if this happens, Lucretius says, there has been a break in the chain of consciousness, and thus death is nothing to us. So, although the hypothetical scenario of this thought experiment is certainly not something that human beings may bring about themselves, Epicurean physics leaves open the possibility of our atoms being reassembled after our death, and hence the thought experiment cannot be said to be built on an impossible scenario. It is, therefore, plausible to infer that the kind of hypothetical assumptions ancient philosophers used in thought experiments, whether possible or impossible, could not be the criterion on the basis of which an argument should or should not be regarded as a thought experiment.

Now, since an ancient thought experiment does not have to be a *reductio ad absurdum* argument, and since it is irrelevant whether the initial assumption on which it is built is possible or impossible, we should next look more closely at the way the hypothetical scenario itself is constructed in Archytas' thought experiment. As we have seen, the positing of the initial hypothesis opens two different possibilities that need to be examined; it may be that the man will or will not be able to extend his hand or his stick, and in the Epicurean version it may be that the spear will follow its trajectory or will bounce back. Indeed, this pattern of the two *prima facie* open possibilities seems to characterise all cases of ancient thought experiments. For instance, in Cicero 's *De officiis* (3.89–90) three Stoic thought experiments are presented which were discussed by Hecaton: (i) Imagine a good man at a time when the corn is extremely dear, would he feed his slave household or not? (ii) Imagine a foolish man seizing a plank after a shipwreck, would the wise man grab it from him or not? (iii) Imagine there are one plank and two wise men, would each of them grab it for himself or would one give in to the other? There are many interesting ethical issues raised in connection with these Stoic thought experiments, but what is important for our purposes here is to note that all cases present us with a dilemma after the positing of the initial assumption. In fact, Cicero explicitly says that Hecaton argued on either side of this dilemma, before he undertook to defend his own view. Also, when Plutarch (*Theseus* 23.1) refers to the well-known thought experiment of Theseus' ship, he calls it "controversial" (*amphidoksoumenon*), since some ancient philosophers argued that the ship which is built after the removal and replacement of the rotten planks of Theseus' ship is not the same as the original ship, while others insisted that it is the same.

It seems, therefore, that we should content ourselves with a rather general characterisation of what an ancient thought experiment was: it was constructed on the basis of a hypothetical scenario starting with an imaginary assumption. If this assumption was the contrary of what the thought experiment was supposed to prove, then the thought experiment had the form of a *reductio ad absurdum* argument; but this was not invariably the case. Also, the imaginary assumption could have been either possible or impossible; when possible, it meant either that human beings could have brought it about or that it was in accordance with some physical principles and laws. However, what seems to be a required condition for an ancient thought experiment is the fact that its imaginary assumption leaves open two different possibilities that one needs to investigate, and it is by no means clear from the beginning, which of the two one should follow. That is to say, the imaginary hypothesis initiates a process of thinking without a previously settled or determined conclusion, namely, a series of arguments that should be clearly spelt out, compelling us to make up our mind on a particular subject.

Thin and vague though it may sound, the general characterisation of an ancient thought experiment as we have proposed above, at least manages to exclude certain arguments that have been treated by modern scholars as thought experiments. For instance, to claim, as has recently been the case, that Thales' hypothesis, according to which the earth floats on water like a log, is the imaginary assumption of a thought experiment meant to prove that the earth remains in its place, is in my opinion misleading.[14] For given such a generous understanding of what an ancient thought experiment was, the notion of a thought experiment threatens to lose whatever explanatory power it may have. And this could prove particularly confusing, especially when applied to the case of antiquity, since there is no Greek term corresponding to what we nowadays refer to as thought experiments. But ancient philosophers did use many thought experiments, and even if they didn't conceive of them as a separate category of arguments, I believe it makes perfect sense to rely on our modern conceptual apparatus in order to closely study and analyse the forms of arguments employed in ancient philosophy as thought experiments.

[14] See Rescher, "Thought Experimentation in Pre-Socratic Philosophy"; Ierodiakonou, "Ancient Thought Experiments."

THOUGHT EXPERIMENTS IN THE *DE ANIMA* COMMENTARIES

Peter Lautner

1. Before discussing the subject referred to in the title, I may have to say a few words on the genre. As the term itself is a modern coinage one might be suspicious about its application to issues in antiquity.[1] What we have to see, I believe, is that there might have been a style of argumentation that fits in with the framework of thought experiments. To find it, the first question to be raised is this: what is the difference between what was later to be called thought experiment and other kinds of philosophical reasoning? First, it seems that thought experiment does not require abstract language, but contains a story. Take Archytas' argument. It is not heavily loaded with abstract terms and has a clear story. If someone gets to the boundary of the universe, he may or may not be able to stretch his stick beyond that limit. The style does not remind of Plato's *Parmenides* or Aristotle's *Metaphysics*. Thus we have a kind of philosophical narrative which differs from direct argumentation. However, there is another type of narrative which was also considered as a philosophical reasoning, the myth. The best examples are given in Plato's dialogues, but we find such kind of narrative in later authors, such as Plotinus, abundantly. Although Plato's myths are for the most part invented, Plotinus could also adopt traditional stories for his own purposes. What, then, is the difference between philosophical myth and thought experiment? One might suggest that the former does not only contain a story but is also hard to translate into straightforward philosophical reasoning, whereas thought experiments, such as Archytas,' are more concise and serve to illustrate a thesis. There is an important borderline case, however, the story about Gyges' ring in Plato's *Republic* (359B6–360D7). It may take the form of a conditional saying that, "if you can avoid all the punishment consequent on your being unjust—by using a ring which makes you invisible—then you would abandon justice for injustice." But we have to make a

[1] For the term, see Brown, "Thought Experiments"; Zalta, *Stanford Encyclopedia of Philosophy*; and the introduction of this volume. For a pioneer analysis of thought experiments in antiquity, see Ierodiakonou, "Ancient Thought Experiments," 125–141.

distinction here. One is the myth about the Lydian boy Gyges finding the ring and thereby advancing his interest ruthlessly to the effect of establishing himself as the king. Another is the question the myth carries over to the dispute about justice. The myth is a straightforward story with a clear end (Gyges becomes king) which does not leave us with a problem to solve.[2] By contrast, thought experiment leaves us with something to explain, even if it is not always aporetic. Even if we are left with a question in the end, the solution is to be worked out and we are not compelled to admit that the problem is intractable in the present state of argument.

Thus we have established three conditions. A thought experiment is not formulated in abstract language, has a story and leaves us with an *explanandum*. We can refine the thesis by saying that they are hypothetical or counterfactual states of affairs.[3] The question is whether this kind of reasoning fits our modern conception of thought experiment. Can we explain our ancient thought experiments in terms of counterfactual reasoning? To give a positive answer we have to rely on the distinction between weak and strong counterfactuals.[4] Strong counterfactuals refer to events or states of affairs that cannot happen. On the other hand, by weak counterfactual we mean statements referring to events that do not occur actually, though it is not inconceivable that they may occur. As a matter of fact, we find in antiquity ample examples of such kind of reasoning that qualifies as thought experiment in the weak sense. It does not rule out, however, that many arguments draw on an inconceivable scenario.

An example of the second kind has been discussed in Katerina Ierodiakonou's paper. It has a strong metaphysical commitment and concerns a Stoic query, the solution of which is not quite clear, anyway. The problem of identity was illuminated by the example of Dion and Theon, the latter being the same as the former except for lacking a leg.[5] What happens if Dion loses the leg that Theon is lacking? One of them has to perish because they will occupy the same portion of space and share the same substrate. On Stoic views, however, two individuals cannot occupy exactly the same place and cannot possess the same qualities, since the principle of the identity of indiscernibles rules it out. Unfortunately, for

[2] See Brisson, "Plato," 141–158, and Virvidakis, "Η Φαινομενολογία," 525–535. On Gyges' ring in this respect, see Wilkes, *Real People*, 5, 11.

[3] See Ierodiakonou, "Ancient Thought Experiments," 131.

[4] The distinction is made in the introduction to this volume.

[5] Referred to in Philo of Alexandria, *De aeternitate mundi* 48 (*SVF* II 397).

lack of textual evidence, it might not be quite clear what function the example has in the whole set of argumentation.[6]

The first kind of thought experiment may be illustrated by Carneades' plank.[7] It is intended to raise a question concerning justified murder. If murder is pardonable only when a person kills another person in order to prevent him to cause his own death then we might face the following problem. Imagine that after shipwreck you find yourself in the sea with a sole plank in sight. The plank can hold one person only, and has already been occupied. It is clear that, if you let it float away you will die and the person sitting on the plank will survive. If you want to live you have to push him off the plank. In other words, in order to live you have to kill him. Otherwise, in a sense, he will be responsible for your death. The question becomes more intricate if we have to keep in mind that the person swimming in the sea is a Stoic wise, whereas the one sitting on the plank is a fool. Or, what if both are wise? Again, it is not clear what purpose this example may have served. It might have been nothing but a test for the initial definition. Whatever the moral of the story may be, however, the narrative depicts an event which is by no means impossible to happen. If we are ready to call this kind of philosophical reasoning thought experiment, even if only in a weak sense, then we are entitled to study certain features in the commentaries from late antiquity under this heading. This is an important prerequisite for the present study.[8]

Thought experiments play an interesting role in Neoplatonic commentaries on the *De anima* written in late antiquity for at least two reasons. The status of psychology was ambivalent since, on the one hand, the commentators inherited an Aristotelian framework according to which psychology is one of the natural sciences, on the other hand, however, their Platonic legacy put the science of the soul above the philosophy of nature. Consequently, as a supernatural entity, the soul can be investigated in various ways, one of them being the examination of the cause through the effects.[9] Against this background, it is interesting to see that the majority

---

[6] It might have been intended as a *reductio ad absurdum* of the Growing Argument, see Sedley, "Stoic criterion," 255–275; Bowin, "Chrysippus' puzzle," 239–251, though the two authors differ in other points.

[7] Mentioned by Cicero, *De officiis* III 89–90 (he does not mention Carneades), and *De re publica* III 30; Lactantius, *Institutiones divinae* V 16. 10.

[8] To find thought experiments of the same kind in modern philosophy one might think of the various versions of the "trolley" introduced by Foot, "The Problem of Abortion."

[9] I have attempted to make some clarifications concerning this issue in "Neoplatonists," 81–109.

of such experiments draw on everyday experience. They are free of counter-intuitive elements. In the first part of the paper I survey the overall feature of these thought experiments in the *De anima* commentaries, whereas the second part is going to be devoted to a case study to illustrate the theoretical output of them. They take the form of examples. Nevertheless, they may be considered thought experiments in the sense that they include a hypothetical scenario which does not occur actually, but there is nothing to prevent it from taking place.

2a. Except for Philoponus, the authors of the commentaries under discussion are unknown, for convenience we may call them Pseudo-Philoponus and Pseudo-Simplicius, all compiling a commentary on Aristotle's work in the 6th century, presumably.[10] Despite some remarkable differences, both doctrinal and theoretical, the commentaries are in agreement that the soul is capable of existing separately from the body, and not subjected to affections.[11] Thus, in discussing the operations of the soul the commentators insist that it cannot admit affections. Even if the intellect has knowledge of the perceptible things, it undergoes a so-called perfective affection, which, in Pseudo-Philoponus' view (*in DA III*, 526.30–34), is better called freedom from affection. With this overall picture in mind, both commentators had to account for the activities of the soul in terms of incorporeal processes. In some cases, when discussing those parts of the human soul that have counterparts in animals, analogical reasoning was allowed.[12] That is, we can use alleged evidence in order to discover animal assent, imagination in worms, or animal capacity to tell the differ-

[10] Pseudo-Philoponus is the author of the Greek commentary on Aristotle's *De anima* III, published in CAG XV: See more recently, William Charlton in the introduction to his translation of Philoponus' *In De anima III*, see *'Philoponus,' On Aristotle On the Soul 3.1–8*, and *'Philoponus,' On Aristotle On the Soul 3.9–13*. He argues that the author is Stephanus. But the question of authorship of the commentary on Book III does not affect the assumption that Philoponus is the author or compiler of the commentary on Book II. The commentary attributed by the manuscripts to Simplicius is also of dubious authorship. The other candidate is Priscian of Lydia, as has been argued by Carlos Steel, "Introduction." The by now classic statement of Priscian's authorship has been made by Steel and Bossier, "Priscianus Lydus," contested in detail by Hadot, "Simplicius or Priscianus?," 159–199. For the most recent case of Priscian's authorship, see Perkams, "Priscian of Lydia," 510–530.

[11] On the *apatheia* of the soul, see Philoponus, *in DA* 70.19–21; 164.2–3; 443.6–8; 521.11–12, 18–29; Pseudo-Philoponus, *in DA III*, 542.4; Pseudo-Simplicius, *in DA* 45.27; 228.14–17.

[12] To mention just one example, the question whether certain animals such as worms, bees and ants have *phantasia* is decided with reference to regularity, see Pseudo-Philoponus *in DA* 495.5–19.

ence between sweet and white, which is a higher order perceptual activity in Aristotle. Evidence is reached by transferring the explanatory scheme of the psychic activities of humans onto animal behaviour.

Given the considerable gap between the material/corporeal and the immaterial/incorporeal, one might expect that our Neoplatonist commentators had to restrict the sources for their arguments concerning the essence and activities of the soul. The description fits the approach of Pseudo-Simplicius particularly well. On the other hand, what we can observe in Philoponus is that he used a great amount of medical material in the discussion of the activity of the soul.[13] He made use of such doctrines when discussing visual and auditive processes. Interestingly, he also used the material, in a critical way, in the analysis of second order perception, in examining the problem of how we perceive that we perceive (*in DA* 293-13-26).

More to the point, some basic methodical principles concerning the study of the soul do not rule out the application of thought experiments that draw on everyday experience. Following the principles laid down in Aristotle's *Physics* I 1, the commentators assume that we can study the causes through effects. In the case of the soul, effects were also the various activities from vegetative processes up to sense perception and *phantasia*. The method proceeds from what is clear to us at perceptual level to what is clear at a natural level. This was very clearly put in Pseudo-Simplicius. First, we have to make a distinction between essence and attribute. Then, among the various attributes, we have to choose those, which are proper to the soul (*in DA* 9.3–4). Sense perception is a most natural choice because it is not only a proper attribute of animals, but also the distinguishing feature that separates animals from plants. If we want to understand the essence of the soul, a perfectly legitimate way to do this is to study activities such as sense perception.

Such a method allows the commentators to get support from everyday experience concerning sense perception or to illustrate the theories with thought experiments produced by the aid of imagination. For this reason, they could use thought experiments in the weak sense. In order to test or prove a case they can put up imaginary scenarios that are by no means impossible. The route taken by Philoponus also illustrates the thesis. Even if his theory of the soul is markedly different from the one held by his Athenian colleague—and adversary—, he insists that bodily activities are

[13] This has been described by Todd, "Philosophy and Medicine," 103–110.

indirect evidence (*tekmêrion*) of the activities of the soul (*in DA* 106.24). The reason is that the different activities of the soul are directed towards their proper objects: desire towards the object of desire, sense perception towards the sensible object. In most cases, Philoponus admits that for this reason investigation into the proper objects precedes investigation into the activity itself. The only exception he allows for is the intellect. The intelligible object transcends us and therefore is more difficult to define, whereas the intellect is a faculty that belongs to us (39.15–24). In sum, the use of thought experiments relying on everyday experience was not irrelevant in the discussion of the nature of the soul. Strong methodical principles dictate that investigation into the nature of the soul recurs to material drawn from common experience. This is not to say that they had to use thought experiments in the weak sense only. But if they did use them, it was because they may not have found any obstacle to revert to everyday experience and devise arguments, including thought experiments, accordingly.

It is important to see that Alexandrian commentators recall everyday experience many times also when using thought experiments in other matters. One of the famous examples serves to illustrate the paradoxical nature of Aristotle's theory of projectile motion. On explaining the reason why a javelin continues to move after it has left the hand that threw it, Aristotle assumed that there must be a cause external to the javelin, which, on his assumption, is nothing but the successive pockets of air immediately behind the javelin. These pockets of air received the power to push it onwards even if the hand had come to rest.[14] The Aristotelian theory had to face quite a few problems. An alternative explanation of projectile motion, which is free from such difficulties, was proposed by Philoponus. He suggested that there is a force that the thrower implanted directly into the javelin. Consequently, the force moving the javelin is not external to it. To show the absurd consequence of Aristotle's theory, Philoponus invites us to conjure up an army. If Aristotle's insistence on the pockets of air were correct, then the army would not have to touch its projectiles. All it would have to do is to perch the projectiles on a thin parapet and set the air behind it in motion with 10,000 pairs of bellows (*in Phys.* 641.13–642.20). According to Aristotle's theory, the projectiles should fly towards the enemy, but in fact they would fall down immediately and not move the distance of a cubit. If we compare this

[14] *Phys.* VIII 10, 267a2–12. For an analysis of the theory itself, see Sorabji, *Matter, Space & Motion*, 227–249.

experiment with those in the *De anima* commentaries, we can see that reliance on everyday experience was not confined to the study of the soul. It was, in fact, a quite common phenomenon. There is, however, an important difference between this example and those, which are to be discussed in the context of the *De anima* commentaries. The discussion of projectile motion makes use of a hypothetical scenario since no one would take seriously the idea of perching javelins on a parapet and setting the air behind it in motion.[15] By contrast, the narratives in the *De anima* commentaries rely on common experience since they refer to events that can happen to everybody. Still, they have a common ground, for none of the scenarios proves to be essentially impossible. They are not like the absurd experiment with the imaginary Theon and Dion. If we construe thought experiments to discuss problems concerning mental processes, it might be fairly difficult to devise counterfactuals that in fact cannot happen. Application of strong counterfactuals would mean that we have to figure up such mental processes that cannot occur at all. This would amount to something like imagining the unimaginable.

2b. In what follows, I shall concentrate on one passage in Aristotle's *De anima* (III 2, 425b12–23) and examine two rival interpretations of the problem exposed there. To support their claim, both commentators supply us with short thought experiments serving their own, opposing purposes. This part of the paper illustrates not only how the two Neoplatonists made use of such a tool, but may also highlight a dispute among the Neoplatonists on the scope and role of sense perception. We shall see that selection of different aspects of basically the same thought experiment had an important role in the argument itself.

In the passage of the *De anima* Aristotle is looking for a factor responsible for perceiving that we are perceiving. The options are as follows: it is either the task of a separate sense or faculty to apprehend the activity of the senses, or the sense itself is capable of apprehending its own activity. Aristotle eliminates the possibility that we are aware that we see by means of another sense. That would lead to the assumption that the object of sight, which is the colour, is an object of this other sense as well. This assumption, however, goes against the core of Aristotle's conception of sense perception. Each particular sense has a specific object, which is not shared with other senses. Sight apprehends colours only, and colour

[15] I owe this point to Katerina Ierodiakonou.

is not apprehended by other senses. Furthermore, if the sense apprehending the activity of sight were different from sight, either there will be an infinite regress or else, to rule it out, there will be a sense or other cognitive faculty, which is concerned with itself. Infinite regress should be avoided by all means, whereas the notion of a sense apprehending its activity seems acceptable, with the qualification that there is no need for a separate sense credited with the job of grasping the activity of sight, along with its own activity. Aristotle suggests us not to move beyond the visual capacity and admits that it can grasp its own activity. The suggestion is not without difficulties. If to perceive by sight is to see colours and that which possesses colour, then the sight apprehending its own activity must possess colour. For this reason, one has to admit that to perceive by sight is not a single thing. Even when we do not see anything, it is by sight that we discriminate both darkness and light, although discrimination of darkness and light is not to be equated with perception of colours. Furthermore, sight itself can also be coloured in a way, for each sense organ receives the object of sense perception without its matter.

Now, if we turn to the Neoplatonist commentators to see how they interpreted the passage, we find wildly different approaches. The only point of agreement is that all of them were keen to avoid infinite regress. Just to remain with the two authors whose texts I shall examine next, they differ not only in the results of the inquiry, but also in their attitude towards Aristotle's views. Whereas Pseudo-Simplicius remains faithful to the principle that Aristotle was basically right and his views are to be reconciled with those of Plato, Pseudo-Philoponus explicitly rejects much of what Aristotle said about sense perception.

Let us start with Pseudo-Simplicius. He explains the passage in *De anima* III 2 with reference to the general theory he takes over from Iamblichus. He shares Iamblichus' view by saying that in the case of human beings the rational soul penetrates the perceptive power so that this latter, too, becomes rational, and what is more, the body itself is also provided with organs in a rational manner.[16] This rational pattern is so decisive that, as a result, human perception and animal perception are only homonyms, that is, they have in common just the name, but according to the definition they are different. In Iamblichus, this doctrine is perfectly embedded in his general theory of the soul according to which different species of animals have different kinds of soul and there-

[16] *In DA* 167.35–188.3.

fore the soul of a non-rational animal can transmigrate only into the body which used to belong to another non-rational animal, and human soul can be reincarnated in a human body alone. But if perception, just like human desire, is rational, it has to possess the attributes of rational faculties and thus its activity has to return to itself. In sense perception, rationality manifests itself in the ability of the sense to reflect on itself. This is the general theory, which Pseudo-Simplicius endorses, but this is not a proof of the specific feature of the sensory capacity, only an application of the theory to a particular case. No doubt, however, that Aristotle's theory supplies him with the necessary conceptual tools. He follows Aristotle's arguments that there must be such a thing as a sense that apprehends itself. But, if so, we have to assign this ability to sight and the other senses as well, since it seems to be nonsense to suppose that a single particular sense could grasp the activities of the rest, too. On resembling reason, the senses strive to assimilate reason in such a way that simultaneously with their perceiving of objects they perceive themselves also, and in this way imitate reason as returning to itself. Moreover, this imitation finds its expression also in the fact that the sense has a finite nature and is connected to what is without parts, which is reason. The capacity for perceptual awareness is due to a simultaneous perception. Most interestingly, this simultaneous perception works even when the objects outside the senses are missing or cannot be perceived. Alluding to the Aristotelian text, the commentator takes the example of seeing in darkness. It is by virtue of this joint perception, he says, that I know that I am not blind and that my seeing faculty is functioning, although because of darkness I do not perceive any object. The proper objects of sight are colours, and darkness does not contain any colour in actuality. Light and darkness are not colours, but causes of visibility.[17] Thus they are prerequisite for seeing colours. In this case, therefore, the activity of the sense consists not in seeing something, say the dark colours, but only in the attempt to see. It can be said that it is even a result of not seeing that sight discerns its own activity when the perceptible objects do not fall under the sense that is attempting to perceive. In the course of this attempt to see, then, what the sense perceives is not the act of seeing but the attempt itself, and thus what the simultaneous perception grasps is not the seeing

[17] Ibid., 129.8–12, 30–35. It is an interesting problem, not pursued in this paper, whether the sense of sight can apprehend light and darkness *per accidens*. It is clear that it cannot perceive them *per se* since they are not colours. In the example of entering a dark room, the emphasis is not on seeing the darkness, but on not seeing anything outside.

but the attempt as missing its object. For this reason, sight also discerns the fact that it does not see anything. Now, in what sense can this example be considered a thought experiment? In the weak sense, I believe, because the example invites us to think about the issue in question. It is not a straightforward philosophical argumentation or a myth. The event it describes is ordinary. But the commentator does not mention it for the sake of mere illustration. He mentions it rather as a contribution to solve a problem concerning perceptual awareness.

Perhaps the commentator has a further aim in using such an example. He accepts Aristotle's view that the primary object of sight is colour and sense perception is nothing but a reception of the form of perceptible things without their matter. To remain with sight, in this view, the presence of colour seems to be the necessary condition for the visual sense to grasp the common sensibles, such as size, magnitude, etc. Colour must also be present when the sense apprehends its own activity. By stressing that sight can work in a reduced way in darkness, the commentator introduces a new thesis. It says that, as darkness and light are not colours, the sense can work even if there is no colour around. Hence the activity of sight does not depend on the presence of colours. The sense of sight functions even if the primary objects of seeing are absent. Thus the commentator ascribes a kind of independence to the particular sense that Aristotle was hesitating to admit. No doubt, the example does not allow us to infer that the sense can work without the presence of anything out there. But it is sufficient to establish a relative autonomy of the senses. This fits in well with Iamblichus' theory, who also emphasised the independence of psychic activities from the physical world. Now the next question is this: What kind of role does the example play here? Is it a mere illustration? Or does it have an argumentative force of its own. It seems that, for the most part, the commentator is content with a remake of what Iamblichus and Aristotle said. The example has to carry the burden of demonstrating that the Aristotelian thesis can be adjusted to Iamblichus' theory. There is no other argument in this part of the commentary, which serves such purposes.

In Pseudo-Philoponus we find a very different approach. He is not very much interested in Iamblichus and his attitude towards Aristotle is not that of a faithful exegete. He rejects the thesis that a particular sense is able to grasp its own activity. He also rejects that perceptual awareness is due to the activity of our sensory power in general. Instead, he endorses the view, shared by some of his contemporaries, that perceptual awareness is due to a special activity of our rational soul. I shall quote the text:

> So Aristotle does not speak rightly but, as we said, it belongs to the attentive part of the soul to get to know the activities of the senses. And that this is so can be seen from the things themselves. For when reason is engrossed with something, even if sight sees, we do not know that it has seen because reason is engrossed. And later, when reason comes to itself and, though not seeing the friend, even now says that it has seen him, it is as if it were taking up a small imprint of the thing seen and, though it was engrossed, now having come to its senses it said that it saw. So it belongs to reason to say 'I saw.'
>
> If anyone raises the difficulty: how can it be said that reason engrossed with universals, but now we say that it acts along with particular senses, say this in reply: when it uses the body as an instrument it is engaged with universals, and now it uses the body.
>
> So much on that. But we are not paying our nursling's dues to Aristotle but making out that his account is false. Let us say, then, that he agrees with us. When he says that the sense knows it is acting he says it grasps this, not as a sense, but as a tool of reason. He said above (425a22–25) that sight knows the sweet not as sight but as common sense, and he says this now in the same way. Not only does he do this, but common usage has this custom, as when it says that the adze made the chair. See! The tool is credited with the work, not that it made it, but the craftsman used it in making it.[18]

Now, we have an example here. It is not exactly about perceptual awareness. Imagine that someone is walking down the street, completely wrapped up in his thoughts. He does not realise what is going on around him. After a while he realises that he had seen a friend of his. For Pseudo-Philoponus, it shows that it belongs to the attentive part of the rational soul to get to know the activity of the sense of sight. The sense works like a wax-tablet containing all sorts of imprints. It receives the form of everything falling within our visual field. Reception of these forms does not require awareness. Although the person saw his friend, he was not aware of seeing him. On the basis of the example, one might say that the sense is purely passive in receiving perceptible forms. If the sense has a purely passive role in this process, there is no reason to assume that it can be active at all. But, if we want to claim that perceptual awareness is due to a certain functioning of the sense, then we must assume that the sense is active, for it must apprehend its own activity. If the sense is purely passive, however, we need another capacity to be aware that we see or hear. In fact, it seems that we also need this capacity to be aware that we see someone or something. Hence, in order to be aware that I see someone now, I

[18] In *DA III*, 466.28–467.12. I follow William Charlton's translation with a slight modification.

have to have two capacities working in tandem: the sense of sight must be receiving the sensible form, while the attentive capacity must be reflecting on the receptive process. The commentator extends the argument on the particular sense to the whole perceptual power and comes to the conclusion that the attentive capacity belongs to the rational soul. From this point of view, the example does not have an independent force. It corroborates the thesis established by other proofs that sense perception is not capable of reverting to itself. From other points of view, however, it cannot be replaced by any other argument in this part of the commentary. First, there is no other argument to show how conscious perception comes about; as we saw, it needs the co-operation of a passive sense and the activity of the attentive power. Second, it is also alone in showing how memory images are formed. It says that we do not need conscious perception in order to form memory images. Our perceptive power works like a camera; it registers everything in our visual field. The images produced in the course of sense perception are stored in our memory. The commentator does not claim that we can recall all of them, but it is sure that some of them can be recalled even if we perceived them without being aware that we perceived *them*. To establish this point the commentator makes use of a hypothetical narrative. As a matter of fact, the narrative is the only ground of the doctrine. It does not illustrate something that has been proved by other means.

It is tempting to contrast Pseudo-Philoponus' theory with Plotinus' account of the formation of memory images. He examines it in detail in *Ennead* IV 4.8, 9–30. Again, I quote the relevant passage:[19]

> When what is perceived makes no difference, or the perception is not at all personally relevant, but is provoked involuntarily by the difference in the things seen, it is only the sense perception which has this experience and the soul does not receive it into its interior, since the difference is not of concern to it either because it meets a need or is of benefit in some other way. And when the soul's activity[20] is directed to other things, and completely directed to them, it will not accept the memory of things like these when they have passed away, since it is not aware of the sense-impression produced by them when they are there. Then again, one might understand the point that things which happen altogether incidentally do not necessarily come to be present in the imaging faculty, and even if they did would not necessarily be there in such a way that it would guard and observe them, but the impression of a thing like this does not productive

[19] I use the translation by A.H. Armstrong.

[20] Armstrong reads ἡ ἐνέργεια αὐτῆς instead of the ἡ ἐνέργεια αὐτή which we have in the *editio minor*.

> a conscious perception, if one took what was said in the following sense. This is what I mean: if it is never a primary consideration to us in local motion to cut through this piece of air and then that, or, even more, to pass through the air at all, we shall not observe the air or have an idea (*ennoia*) of it in our minds as we walk. For if it was not a primary consideration to us to complete a particular stretch of the road, but we could go on our way through the air, it would be no concern of us at what milestone in the land we were, or how much of the way we had covered; and if we did not have to travel for a particular space of time, but only to travel, and referred no other activity to time, we should not remember successive periods of time.

Clearly, Plotinus insists that we have to attend to what we perceive in order to have a memory image of that. If we do not attend to the different pockets of air on the way back home, for instance, we shall never remember them in the future. It is very hard to resist the temptation to claim that Pseudo-Philoponus is responding to Plotinus. Not only the problem to be discussed is common, but the authors support their theses with the same type of narrative. But even if they use similar narratives they use them to point towards different solutions. In order to get the preferable result they emphasise different aspects of the case. Plotinus draws attention to the fact that after having got home I may not remember each milestone I passed by, while Pseudo-Philoponus mentions another possible component, a friend, whom I just walked by without noticing her, though at home I shall remember seeing her. Milestones differ from friends, but both philosophers use the same overall pattern to make their own point. The difference between their solutions arises from deeper divergence, from their diverging views of the nature of sense perception. As for the nature of their argument, if we combine the accounts of Pseudo-Philoponus and Plotinus we have what Katerina Ierodiakonou considers as an ancient thought experiment.[21] In both cases we start from a similar imaginary assumption: while walking on the street someone passes by somebody or something without paying attention to her/it. Two possibilities are open to him when back at home: either recall that he passed by a friend or forget it. Nothing rules out the assumption that Pseudo-Philoponus simply took over Plotinus' narrative and modified it in order to draw another conclusion. All this shows that the "walking home absorbed in thought" scenario offered a general framework that could have been filled in by different authors in different ways.[22]

---

[21] See the paper published in this volume and "Ancient Thought Experiments," 129, where she warns that we should not conflate it with hypothetical reasoning of any type.

[22] For a modern counterpart of this type of thought experiment, see Dretske, *Knowl-*

In sum, there remains the question if these examples and narratives are to be considered thought experiments or not. Previously, I tried to show that if we allow for the possibility of a weak counterfactual, meaning by that a scenario that can happen at some point even if it does not happen right now, then we have thought experiments in the *De anima* commentaries. Of course, we do not have to accept the thesis of weak counterfactuals.[23] But then there is something that calls for explanation. If we stick to strong counterfactuals then it seems that we rule out, not only a type of narrative, but also narratives related to specific branches of philosophy. In antiquity at least, it seems that the nature of thought experiment is determined by the part of philosophy to which it has been applied. In ethics, as has been shown by the example of Carneades' plank, and in the theory of the soul, thought experiment might not take the form of the narratives we have been accustomed to in physics (Archytas' stick) and in metaphysics (e.g., Theon and Dion). They seem to be more down to earth and deprived of the paradoxical features that strike us when reading thought experiments pertaining to physics or metaphysics alike.[24]

---

edge and the Form of Information, 146–147, referring to a case of someone seeing 27 children in the playground. The person may have seen all of them, but he is unaware of how many he saw. He saw 27 children, but this precise numerical information does not reflect what he knows or believes. As a consequence we may say that there is no cognitive representation of this fact.

[23] In the Grenoble meeting, John Zeimbekis suggested that these examples are experiences in thought, related to mental driven processes. He calls them "mental-mental" simulation (see his paper in this volume).

[24] My thanks go to Katerina Ierodiakonou for her many insightful remarks on earlier versions of this paper.

THOUGHT EXPERIMENTS IN LATE MEDIEVAL DEBATES ON ATOMISM

Christophe Grellard

Introduction

As is well-known, medieval natural philosophy, which commentators often write off as "empiricism without observations,"[1] made wide use of thought experiments as a tool of investigation or falsification. From analogies or hypotheses warranted by imagination and God's absolute power, medieval philosophers tried to define the precise realm of validity of Aristotelian concepts, and to modify them from inside.[2] By invoking God's absolute power, these philosophers also managed to reach a level superior to that of common experience. In this context, a thought experiment is best understood as an imaginary scenario whereby empirical concepts are applied to unobservable phenomena. It allows to visualise a quasi-empirical situation, possibly leading to the discovery of new laws of nature. Hence, it has both a heuristic role and a metaphysical meaning since it reaches a stable phenomenon beyond the mutability of accidents. One of these stable but unobservable objects I would like to talk about here is the atom, or point, or indivisible, that is, an extensionless entity which is the ultimate constituent of a body. As is well-known, atomism enjoyed a revival amongst a minority of natural philosophers and theologians in the first part of the 14th century. However, because of their inability to properly resolve some mathematical arguments and thought experiments, these thinkers are not highly thought of by scholars in the history of medieval science and philosophy. Yet, as I would like to argue, this reputation might not be entirely deserved. I would like to examine

[1] Grant, *God and Reason*, 160.

[2] A first approach of the medieval status of thought experiments is given in King, "Mediaeval Thought-Experiments," 43–64. The importance of God's power in scientific knowledge is sometimes related to the condemnation of 1277. This was Duhem's position, challenged by Koyré. See respectively Duhem, *Études*, vol. 2, 411 sqq., and Koyré, *Études d'histoire*, 37–92. The connection between this condemnation and medieval science is still under discussion. On this topic, see Bianchi, *Il vescovo e i filosofi*, 122–132.

two atomist thought experiments and different ways in which their opponents tried to respond to them. By examining these responses, I hope to clarify the epistemic status of thought experiments. Thus, is a thought experiment an argument or an actual experiment? How can an imaginative experiment be verisimilar and bring about new knowledge of nature (assuming it can), and finally, what can it teach us about the medieval conception of science? For this, I will first present two atomist thought experiments, and successively examine two kinds of response.

1. *Two Indivisibilist Thought Experiments*

The main problem for medieval defenders of atoms or indivisibles was how to deal with a natural but imperceptible object. In order to make this entity manifest and to explain its properties, several tools were used: particularly analogies with sensible things, and thought experiments. The aim was to defend the necessity of atoms in the composition of a continuum, and even, as was the case for some of these medieval atomists, to defend the necessity of an infinite number of actual entities in the continuum. Thought experiments seemed to be useful as a way of refuting divisibilism and introducing such novel and unusual objects as points or actual infinity.

1.1. *Henry of Harclay and the Promotion of Indivisibilism*

The Oxford Chancellor Henry of Harclay (ca. 1270–1317) is generally regarded as the originator of medieval atomism. Among his many arguments, I would like to discuss two thought experiments in favour of atomism.

The first one, a *ratio communis* according to a later opponent, examines the contact between a sphere and a plane:

> Again, a sphere moved over a plane touches the plane only at a point. For whether in a plane or a straight line, a circle or a sphere, there is nothing common but a point, and contact is in virtue of some common thing. But a sphere touching a plane in that manner can move continuously over a plane, therefore it can touch the plane continuously, point after point. Consequently such a line as it describes with its motion is composed of points.[3]

[3] This argument from *Henry of Harclay* is quoted by Adam Wodeham *Tractatus de indivisibilibus*. See ibid., q. 1, a. 2, § 4, 95. On Harclay's argument, see Murdoch, "Henry of Harclay and the Infinite," 220–261; Dales "Henry of Harclay on the Infinite," 295–301.

If a sphere is moved on a plane surface, the only common thing between the two figures is a point. The contact will occur on this indivisible point. And, since there can be a local and continuous movement of the sphere on the line, the sphere will touch the line successively in contiguous different points. If one rejects this kind of point, one would have to assume either a deformation of the figures (the sphere becomes a line, or the line adopts a spherical form), or a penetration of the two figures. Both claims are absurd, since we are dealing with ideal figures as God can produce them by his absolute power. Moreover, not only is God able to produce such an ideal sphere, but also he actually does. Quoting Aristotle's *On Heavens* (II 4, 286b10–287b21), Walter Chatton (ca. 1285–1343) claimed in his version of the argument that the heavens constitute such a sphere. Such an appeal to physical fact reinforces the likeness of the experiment:

> The Second objection against this is about the spherical body which touches a plane only at a point. Indeed, if it was tangent at a divisible point, either the first body would not be spherical, or the other one would not be plane. And there is no contradiction if God produces such things. Moreover, the Heavens, according to the Philosopher, actually are such, otherwise there would be a void.[4]

Hence, defenders of indivisibilism can claim to have an experiment (Henry explicitly uses the term *experientia*), which deals with an ideal situation that is both physically and logically possible, and such that God's absolute power could make it real.

In the second thought experiment, recourse to God's absolute power is much more important. Let's consider how God can see a line: necessarily he will actually see and know all the points in the line. Indeed, either God sees the first point and the contiguous point, or there are other unseen mediate points between them:

> God knows and actually sees all the points which are in a continuous line. Since according to the Apostle "all things are open and laid bare to his eyes." From this: God knows the first inchoative point of the line and every other possible point which can be designated in the same line. Either therefore God sees that between the inchoative point of the line and any other point in the same line a point can be interposed, or he does not. If not, then God sees point immediate to point, which is what we proposed to show. If he does see interposable points, then intermediate points could be assigned in the line, and those intermediate points would not be seen by God, which

[4] Chatton, *Reportatio super Sententias*, II, 2, 3, 116, 31–117, 15.

> is impossible. The inference is evident: For according to the case under consideration a line falls between the first point and any other point seen by God in the same line, and consequently there is some intermediate point.[5]

But the last assumption, that is, the existence of unseen points, is impossible since it would be a limitation of God's knowledge. Nicholas of Autrecourt (1300–1369) put forward another version of the same argument: if an infinite power could divide a line from eternity, it would end up constituting a set made up of an infinite number of points.[6] Obviously this kind of thought experiment is a substitute for natural vision that does not have the capacity to grasp an infinitesimal object. Hence, in order to bypass and make up for the fallibility of human vision, we have to imagine what the divine vision of a continuum could be. A divisibilist would claim that an eternal division will never cease. But such an answer could be perceived as a limitation of God's absolute power. Faced with this problem, Harclay and Autrecourt's strategy was to reduce the infinite to the finite. The continuum is infinite for us due to our limitation, but it is finite to God. God can compare two unequal infinite magnitudes, since for Him they are countable. Thus the main goal of medieval atomists is to show that in the light of the *potentia dei absoluta*, it is not impossible to conceive a continuum made of infinite indivisibles.

1.2. *Towards an Atomist Conception of Thought Experiments*

Which conclusions can we draw from these examples regarding the medieval conception of thought experiments? The first conclusion is that, in both examples, the thought experiment is an extreme case relying on a kind of idealisation that is the suppression of sensible accidents or limits. Idealisation, here, means that our experience has no physical defects or limitations. Indeed, the atom as an ideal object is not a would-be-possible, but a real thing our experience is not able to reach. The aim of this procedure is to reach a level superior to that of common experience. The main tool used to do this is God's absolute power. This tool allows us to maximise our ordinary experience and gives a warrant to the work of imagination since it maintains imagination in the realm of the logically

---

[5] Adam Wodeham quoting Harclay, in Wodeham, *Tractatus*, 93.

[6] Autrecourt, *Universal Treatise*, 80: "Now, however, I do grant that, supposing that any magnitude pointed out to sense or imagination is composed of infinite points, nevertheless, if some agent or agents had divided it from eternity, they would finally come to indivisibles."

and physically possible. The first requirement for a thought experiment is to proceed by means of imagination and God's absolute power (which provide a criterion of plausibility).

The second conclusion is that the thought experiments I have examined seem to have different goals. The goal of the first one is to refute the indivisibilist. It tries to exhibit at least one case where we need to posit an actual and real point in order to save the appearances. The second one has rather a prospective or heuristic goal. In order to introduce the notion of actual infinity, understood as an infinite set of actually existing points, it tries to give a measure to the infinite by reducing it to a species of the finite. Here, the argument seems to rely on a kind of analogy: the infinite is for God as the finite for us. The second requirement for a thought experiment is that it should serve either to falsify a theory, or to produce a new object of knowledge.

The third and final conclusion, which is a consequence of the preceding ones, is that the epistemic status of the thought experiment is different in each case. Even if the first one is presented as an experience, it is clearly a counterfactual argument ("if a perfect sphere could touch a perfect plane, then the contact would be a real point"). By contrast, the second thought experiment, although presented as an argument by Harclay's opponents, clearly is not an argument, but rather a kind of metaphysical vision. Hence, the last requirement of a thought experiment is to be a proof, either direct or indirect.

After these remarks, we can now turn our attention to the strategies developed by the opponents to atomism in order to disarm these thought experiments. How can we disarm a thought experiment? It seems there are two main attitudes. The first is that one seeks to reject the thought experiment for logical or physical reasons. The aim is to detect a fallacy in the argument, or to object to the use of imagination and the appeal to God's power. In the latter case, a new interpretation of the facts described in the thought experiment will allow to draw a radically different conclusion. The second attitude tries to limit the effects of the thought experiment by limiting the context where it could be valid.

2. *The Refutation of Thought Experiments*

The most frequent strategy adopted by the opponents to atomism consists either in showing that the thought experiment contains some logical fallacy, or in showing that the premises must be completed by other factors and newly interpreted. Anyway, in both cases the aim is to refute the conclusion of the thought experiment.

2.1. *The Logical Refutation of Thought Experiments*

The first attitude is adopted by Adam Wodeham (d. 1358) and William of Alnwick (d. 1330) against the thought experiment of God's vision of the points in a line:

> But that argument by Harclay is not cogent, since it is formally flawed according to a fallacy of a figure of speech and also according to a fallacy of the consequent. Therefore in sustaining the false opinion which posits such indivisibles, it should be conceded that God sees every point in a given line, and that he sees its first point and that between that point and any other point of the line, some point and some line is interposed. For this is a universal on the one hand, and a singular on the other; nor is any of its singular false. Whence the point and the line which mediately follow the sign posited in the subject supposits merely confusedly. But what is implied, and in a latter deduction supposed already to be conceded—namely, that there is some intermediate point, or some line is intermediate between the first point of the line and any of its points—this assumption, I say, is false. And on this basis I deny that certain incongruities to which the argument leads could be inferred. Nor does the assumption follow from the prior statement, no more than follows "for every man there is a man seeing him, therefore there is some man seeing every man."[7]

The two Franciscan brothers (Wodeham is probably following Alnwick here) reduced the thought experiment to an argument of the form: "if God sees all the points of line, he sees any point of it." But, as they pointed out, the supposition (i.e., the mode of referring) is different in the antecedent and in the consequent. In the antecedent, the supposition of "point" is only confused (*confusa tantum*) and does not allow us to infer anything regarding any particular point, whereas in the consequent, the supposition is determinate. Once it is formulated as an argument, this thought experiment appears to be defective since it is a fallacy *in figura dictionis*. Finally, if we consider *videre* as a modal operator (similar to knowing), there is another fallacy, since we infer a modality in the compounded sense from a modality in the divided sense: "For every point, God sees it, therefore God sees every point." Such an inference is no more valid than: "For every man there is a man seeing him, therefore there is some man seeing every man."

[7] Wodeham, *Tractatus*, q. 1, a. 3, §3, 103–105.

The same logical strategy is used by John Buridan (ca. 1300–1361) against the argument of the sphere:

> Concerning the sphere place on the plane, one answers that the whole sphere touches the whole plane, when one takes "whole" categorematically. But neither the whole sphere, nor any part of the sphere touches the plane, when one takes "whole" syncategorematically. Moreover, no part touches the plane, when we use the terms syncategorematically, except the last part next to the plane. And we want to signify these concepts when we say that it is not the case that it touches the plane as a line, a superficies, a body or a depth. We want to understand that it does not touch the plane as a whole line, a whole superficies, or a whole body, taking "whole" syncategorematically.[8]

His answer is founded on the difference between the categorematic and syncategorematic senses of a term. The sense of the proposition: "a whole sphere touches the whole plane" is a function of the sense of the term "whole." If "whole" is taken as categorematically, that is as synonymous with "having parts," though in the divisibilist sense, we just mean that two divisible wholes are in contact. But, if "whole" is used as syncategorematically, then we mean that each part of the sphere is in contact successively with each part of the plane. Such a claim is impossible according to Buridan, and in this sense the proposition is false. Indeed, since an indivisible has no parts, it cannot touch another indivisible. Here, semantic analysis allows to uncover an implicit flaw in the argument, indeed, a contact is possible only between things having parts. But no point has parts.

Therefore, in both cases, the logical refutation of the thought experiments wants to present them as invalid arguments due to some semantic mistakes.

2.2. *The Physical Refutation of Thought Experiments*

The second strategy consists in pointing to defects in the thought experiments from the point of view of their physical plausibility. This strategy primarily regards the thought experiment as an experiment. By completing this experiment with some new factors, it tries to change its ultimate significance.

This strategy is mainly used against the thought experiment of the sphere. By denying any contact between the sphere and the plane, the thought experiment is rejected. In this perspective, Adam Wodeham

[8] Buridan, *Quaestiones*, f. 97vb.

suggested two new interpretations of the thought experiment. The first one endeavours to show that no contact is possible between two points:

> To the third objection, I quite concede the possibility of a sphere and a plane, but not on that account that a sphere would touch a plane at something simply indivisible. Instead, either it is not possible that a sphere touches a plane, or if it touches a plane, it will not do so precisely at its point, which will be an indivisible thing. For if such a point were annihilated by God, without destroying anything else on that account, which is quite possible for God, a sphere would nevertheless uniformly touch a plane as it does now ... Nor would the withdrawal of such an indivisible point from a sphere (if we granted the existence of such a point is anything) destroy the roundness of the heavy body, therefore if the sphere were nonetheless to touch the plane, it would touch it in the same fashion after all such points were annihilated as when such a point is posited. And yet it is certain that once every such point was annihilated, if it touched the plane, it would touch it at some divisible. Therefore now too, if as a matter of fact, it touches the plane, it touches it at something infinitely divisible, in a manner to be explicated below.[9]

In order to prove such a thesis, Wodeham calls on the absolute power of God. Let us suppose that a perfect sphere and a perfect plane are possible, and let us suppose that the heavy sphere is descending towards the plane. If God with his absolute power destroys the point at which the contact is supposed to occur, this contact will still be possible since the annihilation of one point does not remove the body's roundness. Hence, the contact does not occur on a point but uniformly from one figure to another. The purpose of this new thought experiment is the same as Buridan's semantic argument: contact cannot take place at one point, since being in contact presupposes having parts and extensions. Adam is probably aware that his argument will not persuade his opponent, given that for any medieval atomist, it suffices for there to be contact that two points be in two contiguous *sitae*. Therefore he suggests a second much more physical interpretation of the thought experiment:

> Things do not touch in either of these ways which are so arranged that between them a continuous medium could naturally be received, or at least such a medium could be interposed by God. But this is so in the case under consideration. Let God create the surrounding air or let him effect this descent and joining in air—granted that it is not something which could be effected naturally. But that joining is false, I say, speaking of that sort of joining permitted by such shapes, while they remained unaltered. This once accomplished, the air interposed would still be continuous, although

[9] Wodeham, *Tractatus*, q. 2, a. 3, § 3, 145.

> not everywhere uniformly or equally thick, but, always thinner or more and more tenuous, as we approach the place of contact. Therefore, some continuum would have been interposed, and consequently those objects would not touch. For, however such touch should be described, it seems that every interpretation requires that the touching object and the object touched be immediate to each other in position.[10]

Let us suppose that the sphere is descending towards the plane, through a medium of air. Such an assumption is reasonable, Adam claimed, as such a physical occurrence is possible either naturally or by means of divine power. Once the sphere has completed its downward motion, there will remain a thin and tenuous continuum of air between the two bodies. Therefore, the sphere and the plane won't be in contact. If we don't accept such a continuous medium, then we are led to posit a void. Indeed, God could remove the sphere without any other change. There would then be either direct contact, which is physically impossible, or a void. Both these new interpretations clearly aim at showing that a point has no physical meaning and plays no part in the contact and thus is useless in trying to save the described phenomenon. By entering air and the void into the discussion, Adam tried to give a much more physical interpretation of Harclay's experiment. Clearly, the experiment we are dealing with here is the kind where some aspects are modified in order to discover new properties. The thought experiment functions as a laboratory of sorts (to the extent that such a notion is relevant at all for medieval science).

Buridan adopted the same strategy in a later text:

> And in connection with this, students have occasionally asked whether a sphere placed on a plane would touch it at a point. I reply that as far as the first description is concerned, it must be denied that it touches at a point, because if it touches, it touches at something, and a point is nothing. But as far as the second description is concerned, I put forward this conditional, that if it touches, it touches at a point in such a way that it touches at the sphere's last part, namely, at the last half, at the last hundredth, a the last thousandth, and so on infinitely. And so it touches not only at a single point, but at infinitely many points in such a way that it touches at infinitely many last parts, by dividing the sphere at circles parallel to each other and the plane itself. But then you will ask categorically whether it touches … I say that they do not touch each other: e.g. if you place an actual sphere above an actual plane here in the air, then there will be air between them and between any and every part of them because the air on the right would touch against the air on the left, for if there were indivisible points as some imagine, those volumes of air would be separated only by a single

[10] Wodeham, *Tractatus*, q. 2, a. 3, §6, 147–149.

> indivisible point, which would not separate the parts joined to it. And since there would be nothing indivisible in the sphere or in the plane, and air is between any indivisible of that sphere and of that plane, it follows that there is nothing belonging to the one that is touching something belonging to the other in such a way that there is not some body between them, namely the air.[11]

Even if you posit a point, the two bodies will be separated by a continuum of air when the descending sphere touches the plane, since this point is dimensionless. From a physical point of view, we have no real contact between the two perfect figures. By introducing the condition of air, Adam Wodeham and Buridan wanted to give a physical interpretation of the thought experiment. It allowed them to minimise the imaginary or ideal dimension of the thought experiment, and to underline the physical impossibility of the thought experiment as described by its proponents. Such an attitude illustrates the difficulty medieval natural philosophers (either atomist or divisibilist) had in arriving at a clear conception of the relation between physics and mathematics. This takes us to the last position: limiting the context of validity of the thought experiment.

3. *The Limitation of Thought Experiments*

The second strategy consists not in refuting the thought experiment, but rather in limiting its scope. It is linked with a reflection on the epistemic import of such thought experiments, and especially about the plausibility and the relevance of such a tool.

3.1. *From Ideality to Reality*

After his semantic and physical treatment of Harclay's thought experiments, Adam Wodeham suggested a last interpretation of the contact between a sphere and a plane:

> But then there is the doubt: how, where and in what manner does a sphere touch the plane? Where I maintain first that although it does not by itself touch the plane primarily and exactly, nor does anything belonging to it primarily and exactly touch the plane, as we know from the first conclusion of the preceding doubt, nevertheless it touches a plane in the manner set forth in the second conclusion of the same doubt, namely by itself and with its parts, namely any part of it extending to and reaching the plane

[11] John Buridan, *Quaestiones de Anima* III, 14 in Zupko, *Buridan's Philosophy of Mind*, 155–156.

> … Second, I maintain that a spherical body touches the plane not only with infinitely many such parts of that body arranged in that fashion, but also with infinitely many parts arranged in different position … Finally in regard to the manner in which a sphere touches a plane, and similarly for other examples by analogy, I maintain that although the sphere touches by means of a divisible, it would not touch divisibly but in the manner of a point.[12]

After recalling that a perfect sphere and a perfect plane can't touch each other (there is always an intermediate body between them), he then claimed we can consider them as two wholes, that is, to use Kretzmann's[13] and Zupko's words,[14] as two macro-indivisibles. In this case, we have to understand the nature of contact as a kind of contiguity, where the limits of the two bodies are simultaneously immediate to each other. Wodeham offers several proofs using a proportional division of the sphere *ad infinitum*, such that we will arrive at a point. In other terms, the bodies are divisible *ad infinitum*, but we can represent them as two wholes, and in this case the contact must be *punctualiter*. Although of their own accord neither the sphere nor the plane touch each other, and even though they do have infinitely divisible parts, *secundum modum tangendi*, they have to be conceived *indivisibiliter*. This could be considered a major concession to indivisibilism. But in Wodeham's perspective, it's only a distinction between a physical and a mathematical point of view. He was contrasting a real case (using words as *apud naturam, de facto*) where there is no contact and where the bodies are infinitely divisible, with an ideal, counterfactual case (*si esset*) where we can imagine an indivisible point. Moreover, the concession is not very important, since later in his argumentation Wodeham takes care to limit the value of imagination:

> But incomplex imagination is also of two kinds, namely simple and composite. Simple imagination is that by means of which motion is produced by the common sense—that is, to which simple exterior sensation corresponds. And it is true that everything imaginable in this sense is possible at least by God's omnipotence, for according to Luke: *every word shall be possible to God*, that is everything conceivable or imaginable using this simple concept. But speaking of incomplex composite imagination, it is not true that everything imaginable is possible. For using that composite concept, I can incomplexly imagine an inanimate man or an impossible or prohibited entity, which I consider the same thing. Also using this mode of imagination, I can imagine not only that a point is indivisible, but also that a body

---

[12] Wodeham, *Tractatus*, q. 2, a. 3, § 14–15, 18, 151–153.
[13] Kretzmann, "Adam Wodeham," 381–397.
[14] Zupko, "Nominalism," 158–185.

> with depth is indivisible. And it certainly includes a contradiction to posit the existence of any of these things so imaginable.[15]

On the one hand, incomplex imagination is simple and part of the common sense. According to this meaning, God can do all that is imaginable. On the other hand, incomplex imagination is composite. Such an imagination is used to think the point. But God can't do all that is imaginable in this sense, since many composite imaginations are contradictory, particularly the imagined point. We can imagine the contact between a sphere and a plane on a point, but God can't make such a contact obtainable. Finally, such a thought experiment seems physically useless.

A similar attitude can be found in Buridan's early writings. Dealing with mathematical imagination, Buridan granted that the point could be a useful tool in order to give an ideal measure. But terms such as "point" or "void" do not refer to anything real. All sentences containing them are false in the literal sense:

> This name "point" is imposed in order to signify as if such indivisible things would exist. Hence, this name "point" signifies the same that this sentence "an indivisible thing having a position in a line." And this is a fictive meaning, as the meaning of the name "void." According to this meaning, it would be true to say "no point does exist," as it is true to say "no void does exist." All sentences of this kind are literally false ... Against this claim, you make an objection: all mathematical sentences would be false. We have to answer that a mathematical sentence is not false since it does not affirm what we can imagine is the case. It affirms that if it would the case, then such a conclusion would follow ... They pose the points categorically due to the brevity, but they think hypothetically, and natural philosophy has to examine whether it is or not the case categorically.[16]

Hence, the thought experiment using the notion of points belongs to mathematics, that is a hypothetical or counterfactual science. Sometimes this science can be useful for natural philosophy: for example, the imagination of epicycles is useful in astronomy. However, once a mathematical entity has been imagined, natural philosophy, as a categorical science, still has the task of examining whether or not such an entity can actually

[15] Wodeham, *Tractatus*, q. 2, a. 3, § 26, 159.

[16] Buridan, *Quaestiones super octos libros Physicorum*: IV, 3. See again, ibid.: "It must be known that this name 'point' is a privative name. ... But a magnitude is named a point as long as it is abstracted from any divisibility. For this reason, we say that a body can't touch another one except as a superficies since, when we speak correctly using 'whole' distributively, my hand touches whole the other hand." On this topic, Celeyrette, "La problématique," 86–108.

occur in nature, and whether it is a real entity or not. In the case of the sphere, the real contact occurs through the entire surfaces.

Finally, atomist thought experiments that purport to defend the reality of points are dismissed as purely abstract constructs, useless for the purpose of natural philosophy since they can be of no help in explaining physical phenomena. The fictitious dimension of the thought experiment is assimilated to a kind of arbitrariness.

3.2. *On the Very Utility of Thought Experiments*

Although both, Adam Wodeham and John Buridan tried to dismiss atomist thought experiments as entirely useless, another divisibilist, Nicholas Oresme (1320–1382), held that they can have a real but limited function in natural philosophy. Developing Buridan's considerations on imagination, he claimed that we have to accept the notion of a point in order to save the geometrical appearances.

In his early work, Oresme still considered mathematical imagination not to be enough to answer the hypothesis of the sphere:

> To the fourth argument, about the sphere moved on a plane, one concedes the case. And there is no point here by which the sphere touches the plane, as it is imagined by the mathematician; but the sphere touches the plane indivisibly anywhere.[17]

We have to use an adverbial description of the phenomenon ("the sphere touches the plane indivisibly (*indivisibiliter*)") rather than a nominal one ("the sphere touches the plane on an indivisible point"). Such an adverbial description involves an ontological commitment to *modi rerum*, that is, intermediate degrees of being, and to the *complexe significabile*. However, in a later work, if he still seemed to accept a close solution (points are indivisible accidents), he also added another one drawn from Buridan (points are imaginations):

> To this I say that "to imagine something" differs from "to opine or believe something." Now if the mathematicians were to believe that points exist, I concede that they would believe falsely, and their opinion would be false. Nevertheless an imagination is not properly said to be either true or false. Whence if someone imagines a chimera, he does not because of this imagine a false thing. If however he imagines that this is true,—a chimera exists, then he imagines a false thing ... And if it said that if points do not exist and there is only hypothetical knowledge of them, just as there could also be hypothetical knowledge of a chimera which does not exist, therefore

[17] Oresme, *Quaestiones super octos libros Physicorum*, VI, 3.

> concerning those things only hypothetical propositions can be formulated, the reply is that the case is not similar, because through propositions made about points and lines in geometry we can know the commensuration of things truly existing. From such hypothetical propositions we come to know certain things categorically. An example is the conclusions which are posited in astronomy and other sciences ... But imaginations concerning a chimera would be useless and therefore the case is not similar.[18]

Although there can be no doubt about Buridan's influence here, Oresme takes the issue of the thought experiment's relevance in a new direction. Like Buridan, he concedes that sentences about points are false according to the literal meaning. There is a difference between believing and imagining: the mathematician doesn't believe in point, that is, he doesn't claim that points are real entities (if he did, he would believe so falsely). But imagination allows us to neglect the problem of truth and falsity.[19] Consequently, mathematics is only a hypothetical science. Now, since the point has a status that is similar to that of the chimera, we can ask the following question: what is the difference between the science of points (mathematics) and the science of chimeras? It is necessary here to distinguish between two kinds of "impossibles": some "impossibles" (e.g. the chimera) are such that we infer anything from them; other "impossibles" (e.g. the point, the void) are only possible false (*falsum possible*) since they can't occur naturally, even if we can use them in counterfactual arguments (e.g. "if a point were something, a unique point of the body of the sphere would be tangent to the plane").[20] There lies the main difference between Oresme and Wodeham. Oresme can now conclude

---

[18] Oresme, *Quaestiones de Spera*, q. 2, 46–48.

[19] Ibid., q. 2, 55–57: "As to this, it can be said that this is false by virtue of speech: a sphere is tangent to a plane in a point. And therefore I say the whole sphere is tangent to the plane itself, but because by imagination mathematically a physical part of a line and a surface can not be imagined immediate to the plane itself, therefore the mathematician says that it is tangent in a point by imagination, so that this conditional statement is true: if a point were something, a unique point of the body of the sphere would be tangent to the plane."

[20] Ibid., q. 2, 50: "To this it is said it is impossible that a point and a line exist. And when you say, "therefore, it includes the contradiction," this is denied. Whence it ought to be known that there are some impossibles from which contradictories immediately follow, and concerning such it is understood that "from an impossible anything at all follows." There are other impossibles from which contradictories do not follow ... And such impossibles are imaginable, as is evident in the Fourth book of the Physics. And such an impossible the Commentator, in the first book of *On the Heavens*, calls possible false." The status of impossibles and its links to imagination in Oresme's natural philosophy is carefully examined by H. Hugonnard-Roche in several papers. Hugonnard-Roche, "Modalités et Argumentation," 145–163; idem, "L'hypothétique," 161–177.

that a science of chimeras would be useless, unlike a science of points and lines, which is of great usefulness in natural philosophy. Therefore, atomism provides us with model that is useful in trying to examine some problems of natural philosophy (e.g. the centre of the world or the poles).

Conclusion

Concluding, it appears that thought experiments, understood as an imaginative scenario, aim to make clear evidence either mediately as an argument, or immediately as an experience or rather an intuition. This imaginative scenario must obey a requirement of plausibility, such that it could be warranted by an analogy with a natural situation or by an action of God's absolute power. Hence, in order to refute a thought experiment, its opponents can treat it alternatively as an argument or an experience. If it is taken as an argument, the refutation will consist in exhibiting a logical flaw, or in limiting the value of the counterfactual premises. If it is taken as an experience, the refutation will consist in introducing new conditions in the observational (that is, imagined) premises in order to modify the conclusion. Nevertheless, in all cases, what reveals the use of thought experiments in the debates on atomism, is the difficulty both atomists and anti-atomists had in understanding the relation between natural philosophy and mathematics. Thought experiments are referred to the realm of mathematics, that is, to the realm of pure abstraction, in order to invalidate their physical relevance. The main value of medieval philosophy was to make thought experiments commensurable to the physical reality, but it remained incapable of using them as a simplified model for depicting a complex natural reality.

PART TWO

THE POSSIBILITY OF THOUGHT EXPERIMENTS

THOUGHT EXPERIMENTS AND INDIRECT PROOFS IN AVERROES, AQUINAS, AND BURIDAN

Simo Knuuttila and Taneli Kukkonen

Modern Aristotle scholars have tended to value the seventh book of Aristotle's *Physics* none too highly. Ancient commentators already considered it an early work that was superseded by the eighth book—an approach exemplified by Simplicius' commentary—or else skipped over it outright, as Eudemus apparently did in his *Physics* and Themistius in his paraphrase.[1] In spite of its poor general reputation, the work came to play an important role in the history of philosophy due to the *reductio* proofs in defence of the principle that there must be a first moved mover in any given sequence of movers.

Before the main argument, Aristotle reasons as follows. The distinction between being in motion and being moved pertains to all mobile entities. Let us assume that something is in motion and, since everything that moves is divisible, let us assume that a part of it is then not moving. When the whole moves, all of its parts move, whereas if a part is at rest, the whole does not move; therefore, the whole does not move in this case and it did not move first in its own right (*per se* and primarily), because that which is at rest when something does not move is moved by something.[2] As regards this argument, one of the problems for ancient and medieval commentators was how any of this could apply to the heavens, given that Aristotle himself had characterised celestial rotation as being natural to the heavenly body, *viz.* aether,[3] and hence as belonging to it essentially and (one would think on the basis of the *Posterior Analytics*) necessarily.

The second argument is formulated in a more explicit manner. If there is no first moved mover in a given series of continuous or contiguous movers, then the motion from A to B during a finite time seems to

[1] For the early reception of the seventh book of the *Physics* see Simplicius, *In Aristotelis Physicorum*. 1036–1037; for a contemporary introduction, see Wardy, *The Chain of Change*.

[2] *Physics* VII.1, 241b44–242a49.

[3] See *De caelo* I.2.

demand an infinite chain of simultaneous finite movers related to this motion. This implies that an infinite movement is performed in a finite time, which in Aristotle's view is an obvious impossibility.[4]

Without entering into the problematic details of Aristotle's proofs (which are many), let us attend to a phrase that his followers found important. Referring to the impossibility which is derived from the criticised position, Aristotle wrote as follows: "That the impossibility follows from a hypothesis makes no difference; we took the hypothesis as possible, and nothing impossible should follow from postulating a possibility."[5] While modern commentators typically encounter great difficulties in trying to make sense of what is going on in Chapter 1 of *Physics* VII as a whole,[6] medieval authors found the modal remark in 242b72–243a31 helpful, not only in terms of the immediate context, but for the purposes of explaining Aristotle's argumentative procedures in general. For reasons that will become clear, medieval commentators were very interested in the fact that there are a great number of indirect proofs in Aristotle's natural philosophy, proofs in which Aristotle invoked thought experiments of various kinds, and they also found it curious that these thought experiments often involve hypotheses which contradict principles elsewhere described as necessary. In *Physics* VII.1 Aristotle seems to say that this can be done because the impossibilities in question are assumed to be possible. Averroes in particular found this illuminating.

1. *Ancient Discussions*

Let us return to Aristotle's remark that the impossibility follows from a hypothesis. For Simplicius and Wardy, and a great many commentators besides, the hypothesis in question is the one that states that both "movers and things moved are infinite in number."[7] Averroes, however, interprets the passage differently. In his view, when Aristotle talked about a hypothesis that, though impossible, is treated as possible, he did not refer to the criticised position itself, but rather to the additional hypothesis according to which all those things that are moved form a whole with their movers. Why pursue this line of argument?

---

[4] *Physics* VII.1, 242a49–b72.

[5] *Physics* VII.1, 242b72–243a31.

[6] See Wardy, *The Chain of Change*, 93–120.

[7] Simplicius, *In Aristotelis Physicorum*, 1047.16–18; Wardy, *The Chain of Change*, 107.

Aristotle himself admitted that when developing a reductive proof, it is often necessary to introduce auxiliary premises in order to complete the argument and derive an impossibility from the criticised position. Averroes took it to be an important part of Aristotle's approach that these additions themselves should never be considered impossible, at least not in the context of the reductive proof. This is because Averroes was interested in the inferential structure of Aristotle's indirect proofs, which Aristotle elucidated in his logic, but whose implementation in his natural philosophy seems unclear. According to Aristotle, a "falsehood is always concluded through falsehoods"[8] and "an impossibility only follows from what is impossible".[9] Accordingly, if the aim of the reductive argument is to prove the impossibility of some specific proposition singled out for critical investigation, it is important that the impossible conclusion of the thought experiment derives from just that proposition and no other. The easiest way to safeguard this would be to make sure that there are no other impossible premises in the argument as a whole. (It is, however, admissible to include one that is false in actuality.) One example of Aristotle's description of his approach is as follows:

> You must also look not only whether something impossible immediately follows from the thesis, but also whether it can result from a hypothesis, as for example, when it is asserted that empty and full of air are the same thing. For it is obvious that if the air has been expelled, the space will not be less empty but emptier, though it will no more be full of air. So by a hypothesis, whether false or true (for it makes no difference), one of these is subverted, and so they cannot be the same.
>
> (Aristotle, *Topics* VII.1, 152b17–24)

Averroes's contention seems problematic, however, given that Aristotle's own added auxiliary premises often do take the form of propositions which appear to run counter to acknowledged Aristotelian necessities, and thus to be impossible. For example, in *Physics* VII.1, Aristotle advanced the hypothesis that anything in motion can also be at rest;[10] in another passage he suggested that any finite being can be conceived of as being greater than it is,[11] and in a third context he expressly stated that the speed of any motion can be greater than it is;[12] in yet a fourth

[8] *Topics* VIII.12, 162b13–14.

[9] *De caelo* I.12, 281b15–16.

[10] Contrary to what is said of the heavens in *De caelo* II.6.

[11] *Physics* III.8, 208a16–19; however, the universe cannot, see *De caelo* I.5–7.

[12] *Physics* VI.2, 232b21–22—but the diurnal motion of the heavens cannot, according to what is said in *De caelo* II.4, 287a23–30; 2.6, 289a1–2.

that the inside part of the universe could be destroyed,[13] and so on. Even in the above example, the hypothesis of the vacuum is something that Aristotle characterised as impossible elsewhere.[14] Indeed, it is striking how many arguments with impossible premises there are in Aristotle's reductive proofs, and it is equally striking that Aristotle never explained their status—unless, that is, we accept the Averroist reading of the modest remark in *Physics* VII.1. According to such a reading, these impossibilities can be assumed to be possible in some sense. Even if this is correct, it remains unclear on the basis of Aristotle's text how exactly something impossible can be taken as possible.[15]

This lacuna gave rise to discussions among commentators before Averroes's times. It appears that the cataloguing of impossible premises in Aristotle's indirect proofs took its start from Galen's criticism of Aristotle's argument in *Physics* VII.1, and from Alexander of Aphrodisias' subsequent defence of Aristotle. No wonder that Aristotle could derive impossible conclusions from other people's opinions, Galen said, seeing how he added into the arguments impossible premises of his own. Specifically, Aristotle's refutation of the Platonic notion of self-motion fails, seeing as his own hypothesis of a part being at rest while the whole moves constitutes an impossible thought experiment. Alexander of Aphrodisias, who disputed with Galen on a number of issues, rose to Aristotle's defence on this point, and in the process drew attention to the fact that there are other examples of Aristotle assuming seeming impossibilities in his natural philosophy. Alexander also wanted to derive some kind of systematic lesson from Aristotle's practices, though it is not entirely clear what that may have been. Alexander's original treatise *Against Galen on Motion* is lost in the Greek; Alexander's explanations are, however, described by Simplicius, and the treatise exists in an Arabic translation. Based on these sources, Alexander's idea seems to have been that the additional impossible premises in Aristotle's reduction arguments are not strict impossibilities, but weaker ones in some sense.

[13] *De caelo* IV.5, 312a22–b19.

[14] *Physics* IV.6–9.

[15] In *Posterior Analytics* I.6 Aristotle said that certain predicates may belong to their subjects at all times without belonging to them necessarily. Some commentators (e.g., van Rijen, *Aspects*) have taken this to mean that Aristotle operated with a distinction between strong essential necessities and weak necessities in the sense of non-essential invariances, which he often treated as necessities. Inseparable accidents would belong to the latter group as well. Whatever the merits of this interpretation, Aristotle did not refer to these modes in connection with the indirect proofs he presents in his natural philosophy.

This is not very helpful as such, nor is the matter made any clearer by Alexander's chosen example of a stronger impossibility, "sailing through rock." The impossibility of two things occupying the same physical place was of course a Peripatetic commonplace, but the way it compares to other assorted impossibilities is left unexplained. This discussion continued in the 6th century in Simplicius' and Philoponus' commentaries and others, and was known to the Arabic philosophers.[16]

Aristotle also operated with impossible premises in other contexts, most notably when studying things not separable in actuality but nonetheless "separated in thought," as his own words have it. A classic example comes from *Metaphysics*,[17] where Aristotle suggested stripping away in thought every feature from a bodily substance until only matter is left, disregarding the principle that matter is nowhere actually found without form. This approach was expanded on by some late ancient thinkers and sometimes called the Eudemian procedure. In a Eudemian procedure, something impossible is assumed in order to see what follows. An example of this comes from John Philoponus:

> It is not absurd to use hypotheses even if they are impossible, for the examination of other things; on this ground, since some people explained the stationary position of the earth by the motion of the heaven, Aristotle says that we should stop the heaven in thought, and ask where the earth will go by nature, so as to see whether the movement of the heavens is really the cause of the stationary position of the earth. (*In Phys.* 575.2–7)

Counterfactual hypotheses of this kind were not uncommon in late ancient philosophy. As abstract constructions, they were not regarded as formulations of possibilities and were commonly called impossible hypotheses.[18] What appears to be lacking is an account of how the thought experiment is supposed to work, given that it is acknowledged to be impossible.

2. *Averroes*

From this we may proceed to how Averroes dealt with Aristotle's arguments with impossible premises. In addition to his commentaries on Aristotle's *Physics*, Averroes wrote nine small independent treatises on physics, several of which pertain to questions concerning the first

[16] See Kukkonen, "Alternatives to Alternatives," 141–153.
[17] *Metaphysics* VII.3, 1029a16–17.
[18] See Martin, "Non-Reductive Arguments."

mover.[19] Indirect proofs receive particular attention in the eighth question of this collection, where Averroes said about the so-called infinite power argument of *Physics* VIII.10 that it appears to be structurally unsound.[20] According to Averroes,

> A certain group of men disputed with Aristotle about the force of this kind of demonstration. They maintained that it is a corrupt demonstration because the possible propositions that Aristotle used are impossible and false with respect to that which he sought to demonstrate by means of those propositions.[21]

In answering this criticism, Averroes said first that it is evident from what is said in the *Prior Analytics* that an impossible conclusion cannot follow from a possible premise, for if there is a necessary connection between two things to the effect that if the first exists the second must exist, it is obvious that if the first is possible the second cannot be impossible.[22] Aristotle mentioned this basic principle of propositional modal logic in *Prior Analytics* I.15, although he did not develop propositional modal logic apart from modal syllogistics. Averroes added that if the consequent is impossible, then the antecedent is impossible as well. Another principle relevant in this context is the description of possibility in *Prior Analytics*[23] as that which, when assumed as actual, does not result in anything impossible.

On the basis of this, Averroes argued that when Aristotle operates with impossible premises of his own in indirect proofs, these are usually propositions that are impossible *per accidens* and possible essentially or *per se*. The conclusions of indirect proofs, meanwhile, are impossible *per se*, and therefore they do not depend on Aristotle's additional premises. Indirect proofs are syllogisms, having as one premise the proposition, which is purported to be false and is in fact impossible *per se*, while the second, innocent premise is often reached through further inferences that may or may not include premises that are impossible *per accidens*. The important thing is that the second premise does not derive from a *per se* impossibility, for when the conclusion is impossible *per se*, it must be so

---

[19] These are translated with explanations by Helen Tunik Goldstein in *Averroes' Questions in Physics*.

[20] For a more detailed analysis see Kukkonen, "The Impossible."

[21] *Questions in Physics* VIII.6, trans. Goldstein, reading "impossible and false" instead of Goldstein's "impossibly false."

[22] *Questions* VIII.8.

[23] *Prior Analytics* I.13, 32a18–20.

because of a premise which is similarly impossible *per se*. Even impossible propositions which are so *per accidens*, are innocent in this regard:

> It is evident that in the case of a syllogism in which both premises are possible, no impossible conclusion whatever can be inferred. If there is, however, only one possible premise in the syllogism, an impossible conclusion cannot, indeed, be inferred from the possible premises. Rather, if such a conclusion is inferred, it must follow from the impossible premise in the syllogism. It makes no difference whether the possible premise is possible essentially, but impossible accidentally (e.g., in a certain place), or whether it is possible and in no sense impossible. By this I mean to say that an absolutely impossible conclusion cannot be inferred from the premise which is essentially possible by accidentally impossible.
>
> (Averroes, *Questions in Physics* VIII.8)

Assuming something accidentally impossible, if it is taken not as impossible but as possible causes no harm:

> And since this is so, when the preceding exposition posited something as possible and then an impossibility followed from it, is was essentially possible but accidentally impossible. It was only posited insofar as it was possible (*min jihhati mā huwa mumkin*), not in so far as it was impossible, and in this way we took it together with the premise that was doubted in an argument *per impossibile* (*qiyās al-khalf*), so that the impossible followed from it. But when what is possible is assumed to exist, an impossibility does not follow from it. [Aristotle] uses many proofs resembling this one.
>
> (Averroes, *Compendium of the Physics* VI, 93.1–6)

The distinction employed here plays an important role in Averroes' logic. While his paraphrase of the *Prior Analytics* follows the main lines of Aristotle's text, the separate treatise on modality, *Eight Questions on the Prior Analytics*, evokes new systematic ideas, mainly the theory of accidental and *per se* necessary propositions. In Averroes' modal syllogistic, the necessity premises are restricted to *per se* necessary propositions with *per se* necessary terms. The *per se* necessity of these propositions corresponds to the essential structures in the nature of things; their necessary terms, such as "animal," invariably signify the necessary constituents of things. Accidental terms such as "walking" or "white" signify accidental features. The background to these distinctions is *An. Post.* I.4.[24] Since modal premises are taken to be of the divided type, assertoric premises

[24] Averroes's separate essay on modal syllogistics was translated into Latin as *Quaesita octo in librum Priorum Analyticorum Aristotelis*, which is included in vol. I.2b of Averroes, *Aristotelis opera cum Averrois commentariis*: see IV.3, 83–84. In his paraphrase of the *Prior Analytics* (Latin in ibid., II.1; Arabic in Averroes, *Talkhīṣ kitāb al-qiyās*), Averroes simply says that necessary predications express essential connections; see 45va.

in first figure mixed necessary-assertoric-syllogisms must have a middle term, which in fact is necessary. Universal and particular propositions of this kind are called assertoric *per se* and necessary *per accidens*, since the predicate always belongs to the subject when the subject is actual, even though the invariant connection is not based on the natures of things. Propositions with a necessary subject term and a non-necessary predicate term are said to be temporally assertoric, since the predicate belongs to the subject merely at some times. The first premise in first figure mixed assertoric-necessary syllogisms may be of this kind. The conclusion in the first case is necessary *per accidens* and in the second case temporally assertoric.[25] This is a speculative explanation of Aristotle's asymmetric treatment of mixed necessary-assertoric syllogisms and mixed assertoric-necessary syllogisms.[26]

The practical import of these principles to the matter at hand is as follows. Propositions which are impossible *per accidens* but possible *per se* are false about the existing world and impossible in the sense that they are unchangeably false. (The heavens, for example, never change velocity according to the precepts of Aristotelian physics, and therefore any assumption in a thought experiment of them accelerating or decelerating is invariably at odds with reality, and the corresponding proposition false whenever it is asserted.) They are not essentially impossible, however, because the nature of things does not demand their falsity. Examples of such cases include the possibility of a motion faster than the diurnal motion, the possibility that any body is at rest, the possibility of a body greater than the universe, and so on—all cases discussed by Averroes in his *Questions in Physics*.[27] All these are impossibilities in the actual world, but when they are used in arguments, they are not used as impossibilities, but in so far as they are possible. Their invocation at this point shows how Averroes attempted to make sense of Aristotle's practice in devising thought experiments by means of his own reconstruction of Aristotelian logic.

[25] *Quaesita octo in librum Priorum Analyticorum*, IV.3, 84. See also Thom, *Medieval Modal Systems*, 81–85.

[26] Analogous essentialist ideas were developed in 13thcentury Latin commentaries on modal syllogistics: see Lagerlund, *Modal Syllogistics*, 19–57; Thom, *Medieval Modal Systems*, 93–96; Thom, *Logic and Ontology*, 147–177; Knuuttila, "Medieval Modal Theories," 538–548.

[27] See *Question* 8.10–12 and further examples in Averroes, *Questions in Physics*, 138–141.

In explaining these principles, Averroes said that the notion of body does not define that it is of a certain size, nor does the notion of motion delimit it to a certain speed. Therefore, when Aristotle assumed that there is a body larger than the heavens, this is possible with reference to a body *qua* body, but accidentally impossible with reference to the heavenly body; similarly, when he assumed that for every motion there exists a faster motion, this is essentially possible but accidentally impossible, because no motion faster than the diurnal motion actually exists. For the same reason, Aristotle could (coming back now to the old problem of how to interpret *Physics* VII.1) suppose with respect to anything in motion that a part of it comes to rest, even though in the case of the heavens this will never happen.

> We say that if a moved body somewhere is prohibited from resting (*al-sukūn mumtanaʿ ʿalay-hi*), it is not prohibited from doing so insofar as it is moved, but insofar as it is moved in a specific fashion—I [might] say, for instance, 'insofar as its mover is eternal' or 'insofar as it has no contrary.' As far as its being moved is concerned, however, it could be at rest. Thus the possible here is assumed insofar as it is possible, not insofar as it is impossible, and so the indicated proof follows from it. Many [proofs] employed in mathematics are similar to this one, and their deployment produces no impossibilities there.[28]

The solution seems elegant enough; still, one may wonder what it means that these possibilities can be assumed as actual without any impossibilities following from them. After all, it is not possible to assume them in the actual world or in any other world exhibiting the structure dictated by the precepts of Aristotelian natural philosophy. Things are made more complicated by the fact that the problematic possibilities are combined with strictly impossible propositions.

It is not altogether easy to determine what Averroes meant, when he talked about a philosophical investigator positing (*waḍaʿa*) the impossible as being possible. From other contexts, it becomes at least plain that for Averroes any psychological aspect played little or no role in the process. Although Averroes had a well-developed theory of the imagination and the intellect and the way the two relate to one another, he was clear on the fact that when it comes to resolving questions of

[28] *Compendium of the Physics* VII, 116.7–12; similarly the *Commentary on the Physics* VII, comm. 2, preserved in Latin *In Aristotelis de physico audito libri octo cum Averrois Cordubensis variis in eosdem commentariis*, in *Aristotelis opera cum Averrois commentariis*, IV, 307^v–308^r. Cf. also the comments in Kukkonen, "Alternatives to Alternatives," 155–157.

conceivability and possibility, the natural faculty of imagination, in specific, will be of limited use. Many things the imagination is unable to represent can nonetheless be shown by reason to be possible or even necessary; conversely, many postulates which science proves to be impossible are nonetheless first present to the untrained imagination as perfectly imaginable scenarios.[29] The fact that, e.g., a phantom entity such as body *qua* body is scarcely imaginable should steer us away from assuming that Averroes had anything like conceivability (in the psychological sense) in mind when he set up his analysis of Aristotelian thought experiments. Rather, for him, the thought experiments conducted by the philosopher appeared to operate entirely on the level of logic.

Let us take a look at Averroes' treatment of the standard Aristotelian doctrine that all extensions, and by association all extended things, are infinitely divisible. One may wonder how this is compatible with his own view that the elements, which stand at the lowest level of actuality in the order of nature, have a minimal size. If you take a minimally small piece of water, for example, it has a fixed size.[30] How then is it divisible? Averroes thought that while the volume of water is not divisible *qua* water, it is divisible *qua* being a volume without qualification. Such a division cannot be assumed to take place in the actual world; therefore, it is assumed in the same way as abstract mathematical entities in general are assumed as actual. In operating with counterfactual and in fact counterpossible principles, Averroes did not take these as descriptions of possible states of affairs. The principles are not meant to be partial descriptions of possible worlds. They are possible in some rather abstract sense, even as all attempts to apply them specifically turn out to be impossible.

> If so, the possible in this syllogism is only assumed qua possible, not qua impossible, i.e. the heavy here is only assumed to be susceptible of division because it is heavy, not because it is fire or earth. When the impossible is assumed qua possible, no impossibility results from assuming it. When in this preceding syllogism a possible proposition was assumed *qua* possible, namely, that the heavy *qua* heavy is susceptible of division and a proposition of undoubted truth was joined to it, namely that the point is not susceptible of division, it followed that the point is not heavy; nor is it light. If it were either heavy or light, an impossibility would follow from this, i.e. that the point would be susceptible of division, and this impossibility does not occur because the possible was assumed qua possible. Thus, the only remaining alternative is that, in this discourse, it results from our assum-

[29] See Kukkonen, "Mind and Modal Judgement."
[30] On this peculiar Averroist doctrine, see Glasner, "Ibn Rushd's Theory."

> ing that the point is either heavy or light. Proofs similar to these are what Aristotle frequently made use of in this science.[31]

As we shall see, Averroes' abstractionist approach (as we have come to call the strategy on the basis of Aquinas' later adaptation) became quite popular in the Latin scholastics' treatment of thought experiments and counterfactual possibilities.[32] In the light of this it is important to notice that the approach already came under criticism in Arabic philosophy. Most prominently, Avicenna, in the part corresponding to the *Physics* in his philosophical encyclopaedia *The Healing*, observed that there is something strange in saying that a human being can fly in so far as a human being is an animal. After all, in nature one only ever encounters pigs, horses, and cuckoos, never animals as such: every creature which enjoys extramental existence is a member of a proximate species, that is to say, it is fully specified according to Porphyry's tree. Thus, pigs cannot fly, pure and simple, since a pig-as-animal does not exist in any robust, extramental sense of the word.[33]

Avicenna's objection, which for obvious reasons cannot be aimed at Averroes,[34] must hark back to a different and earlier discussion. One likely candidate is the commentary tradition on Aristotle's *De anima* I.1, where Aristotle offhandedly remarked that the living thing as such is "either nothing, or posterior."[35] In Themistius we find this explained in a manner that generalises from the example about the *zôon* to the relation holding between genus and species in general: the genus either does not exist at all or is a mind-dependent entity, an abstraction without a corresponding entity in the world independent of the soul.[36] That Avicenna knew Themistius' *De anima* paraphrase, in two different Arabic versions no less, is a virtual certainty;[37] he may also have read about Themistius' view from other ancient sources. The author of Pseudo-Simplicius' commentary on *De anima* puts forward the exact same model without objection, for instance (genera are derivative, partial, and mind-dependent

[31] *Paraphrase of On the Heavens* III, 3.1, 289–290 al-ʿAlawī (trans. Goldstein, 134–135).

[32] The notion of Aristotle's proofs proceeding *per abstractionem* is advanced by Aquinas in his late comments on Aristotle's *On the Heavens: In De caelo* III.3.

[33] Avicenna, *Al-Shifāʾ*, 88–89.

[34] Though Aquinas seems to think otherwise; see *In octo libros*, VII.1, 888.

[35] *De anima* I.1, 402b7–9.

[36] Themistius, *In De an.* 3.31–4.11; for remarks see Sirridge, "The Living Thing," 53–55.

[37] See Lyons, *The Arabic Version*, xiii–xiv. Lyons' edition of the Arabic translation is based on a manuscript, in which this passage—in fact, a good portion of the paraphrase of the first book—is missing.

entities), although Philoponus, speaking in the name of Ammonius' school, rejects it.[38]

What these discussions reveal is that the question of counterfactual hypotheses is tied up with some rather fundamental questions regarding the fundamentals of metaphysics. For Avicenna, any question regarding the ontological status of quiddities would be particularly acute, given that he had troubles of his own regarding the issue.[39] Still, because the genus for him is an abstraction and late-born in the order of things, he could feel confident in claiming that the properties of genera cannot be used in counterfactual arguments of the sort that Averroes proposed. What prompted Averroes to claim otherwise is something of a mystery, given that the latter thinker is even more wedded to the notion that the essence and its being—or, put another way, the existent being and its essence—are one and the same.[40] It appears that to Averroes, the abstraction approach was doing too much useful work in explaining Aristotle's thought experiments in natural philosophy for its awkward metaphysical dimension to act as a deterrent.

3. *Aquinas and Buridan*

Let us see how Thomas Aquinas, following Averroes, applied the theory of impossible hypotheses to Aristotle's indirect proofs in the *Physics*. In explaining why Aristotle says that the elements can be continuous with each other and with the celestial bodies, Aquinas wrote as follows:

> But it must be said that the contingent and the impossible are taken in different ways when something is demonstrated of a genus and when something is demonstrated of a species. For when a species is treated, that must be taken as impossible which is repugnant to either the genus or the specific difference from which the nature of the species is constituted. But when a genus is treated, everything which is not repugnant to the nature

---

[38] Pseudo-Simplicius, *In De an.* 13.10–21; Philoponus, *In De an.* 37.17–38.16. Philoponus' objections are based on the dubious premise that Aristotle, too, would have accepted the standard Athenian teaching according to which the genus is actually more real than the species and productive of it it. For this doctrine see, e.g., Proclus, *Procli philosophi Platonici*, 650.

[39] Briefly, Avicenna's essence-existence distinction together with his adoption of the earlier theological concept of "thing" (*shayy'* or the Latin *res*) prompts a question about whether the quiddity (*māhiyya*) of a thing is independent and perhaps even somehow prior to the existent thing. See Wisnovsky, "Notes."

[40] See, e.g., Averroes' criticism of Avicenna on this point in the Fifth Discussion of the *Incoherence of the Incoherence*; also *In Met.* VII, *passim*.

> of the genus is taken as contingent, even though it may be repugnant to the differentia which constitutes the species. For example, if I am speaking of animal, I can say that every animal being winged is contingent. But if I descend to the consideration of man, it would be impossible for this animal to be winged. Now Aristotle is speaking here about movers and mobile objects in general ... Therefore, he states as a contingency that all mobile objects are continuous with each other. This, however, is impossible if mobile objects are considered according to their determinate natures.[41]

Aquinas assumed that the possibilities of a thing are determined by its genus, species, and matter. Something that is possible for it as a member of a genus can be impossible for it as a member of a species. The same holds of its being a member of a species and an individuated being. The possibilities of a moving body, for example, are the possibilities attributed to a subject on the basis of what is not excluded by the concepts of body or motion as abstracted from singular beings. Such possibilities can be accidental impossibilities with respect to real concrete things, but a thought experiment which proceeds *per abstractionem*—this is Aquinas's term—may bypass such considerations, since these modal notions refer to different levels of analysis.[42] Aquinas's formulations illustrate a fairly baroque doctrine of abstract possibilities, levels of possibility associated with corresponding levels of analysis outlined in Porphyry's theory of the predicables. They represent 13th century attempts to develop tools for an Aristotelian-Averroistic counterfactual analysis, one that would operate without the idea of simultaneous alternative domains.

The latter was first introduced through the new modal metaphysics which John Duns Scotus developed in his theological works and which, taken as a whole, represents an important late medieval innovation. The central new idea was to define logical possibility as a basic notion that refers to whatever can be imagined as actual without a contradiction (at a certain moment of time). These statements need not necessarily refer to the historical world we inhabit, but may instead find reference in alternative states of affairs, which are to be conceptualised with the help of the notion of compossibility. The idea of simultaneous alternatives as the basic concept is put forward by Duns Scotus as follows: "I do not call something contingent because it is not always or necessarily the case, but

[41] *In octo libros*, VII.2, 896, translated by R.J. Blackwell et al., Aquinas repeats the same idea in many places: see ibid., IV.12, 538; VI.3, 774; VII. 2, 896; VIII.5, 1009.

[42] *In De caelo* III.3, 4–5; cf. *Summa theologiae* I.85.1, ad 1.

because its opposite could be actual at the very moment when it occurs."[43] The differences with respect to 13th century modal semantics were the refutation of the Aristotelian principle of the necessity of the present and the related association of necessity with omnitemporality, as well as separating modal syllogistic from various essentialist principles which were central in Averroes and in Latin works which often followed Kilwardby's commentary on the *Prior Analytics*. Aristotle's modal syllogistic was regarded as a fragmentary theory and contrary to earlier approaches. William Ockham, John Buridan, and their followers did not try to reconstruct a uniform reading of it, believing that such a reconstruction was not possible.[44]

The new approach to the modalities was extended also to the scholastic interpretation of Aristotle's indirect proofs. For example, John Buridan's discussion of *Physics* VII.1 is based on the idea that Aristotle was capable of drawing a distinction between supernatural and natural modalities, the former of which is to be equated with logical possibility and God's omnipotence, and that when he makes use of a premise which is characterised as impossible, it is to be regarded as impossible in physics, but not in the absolute sense of the word (*simpliciter*), i.e. not as a logical contradiction or a limitation on God's absolute power. In line with this rather dubious historical claim, Buridan stated that any additional premises Aristotle made in his indirect proofs which appear to be impossible (the heavens could stand still, the spheres could be continuous or contiguous) cannot be impossible in the sense that they would be self-contradictory.[45] As evidence for this interpretation, Buridan gave a list of several examples of indirect proofs where Aristotle in his opinion consciously operated with just such a distinction. Of the examples mentioned, the assumption that the dimensions could be something different from the dimensions of bodies, which Aristotle declared impossible in many places,[46] is extensively discussed in question 15 of the first book of Buridan's *Questions on Aristotle's De caelo*.[47] The subject of the question is Aristotle's attempt to demonstrate that an infinite body cannot revolve in a circle (*De caelo* I.5). In his *Questions*, Buridan stated that Aristotle put forward several

[43] Duns Scotus, *De primo principio* IV.4; for Scotus' modal thought, see also Knuuttila, "Modality as Alternativeness."

[44] See Knuuttila, "Medieval Modal Theories," 549–559.

[45] *Quaestiones super octo Physicorum libros Aristotelis*, fol. 104ra–105ra. For Buridan's discussion, see also Knuuttila, "Necessities."

[46] E.g. *Met.* III.2, 998a7–19, *Phys.* III.4, 211b14–25.

[47] *Expositio et quaestiones in Aristotelis De caelo*, 304–311.

reductive arguments in which he assumed that there could be a space distinct from an extended body. In answering the criticism that this should be impossible, and from impossibility anything follows, Buridan stated that Aristotle regarded this merely as a natural impossibility and took it as supernaturally and simply possible.[48] These notions were related in Buridan because merely logical possibilities were realisable possibilities with respect to divine omnipotence, natural possibilities being realisable by natural powers. Buridan employed the same distinction to Aristotle's assumption in *Phys.* IV.8, 215a24–216a12, that there are arbitrarily subtle bodies, and the assumption in *Phys.* VI.2, 232b21–22, that every moving thing can move faster and slower.[49]

Buridan thought that the distinction between natural and supernatural modalities was embedded in Aristotle's arguments. His evidence for this is that otherwise Aristotle's proofs would not demonstrate anything.[50] Referring to divine omnipotence, Buridan told that he was very happy, when he found that "Aristotle to a great extent agreed with our true faith." Being well aware of the interpretation of Averroes and Aquinas, he stated that it is simply wrong:

> The commentator and many others answer to this question by stating that even though it is repugnant to heavenly bodies to be continuous, with respect to their special concept, it is not repugnant with respect to the common concept of moved and moving bodies. As far as Aristotle dealt with things moving and moved without descending to their special concept and keeping to their common concept, he could consider as possible whatever was not repugnant with respect to the common concept of movers or moved things or bodies ... Were this a good solution, one could say as well: every body is at rest, but some bodies are in motion, namely the heavenly bodies; therefore that which is in rest is in motion ..., the second premise is true and the first premise is not repugnant to the concept of motion, rest and body. Therefore this mode of argument is totally invalid. (*In Phys.* VII.2, 105rb)

Buridan thought that when something is possible, it must be able to exist. Investigating possibilities is to think about them as actualised in a coherent context of compossibilities. If something cannot be treated in this way, calling it a possibility is based on a conceptual confusion. This was a problem with Averroes and Aquinas, but not with Aristotle:

---

48 Ibid., 307.

49 Buridan, in these places (IV.9, 76rb, IV.10, 77va, VI.9, 101vb–102ra), did not stress that Aristotle consciously operated with the modal distinction, but he referred to them in order to support this claim in *In Phys.* VII.3.

50 "Aliter nihil valeret eius ratio," see VII.3, 105rb.

> It seems to me that Aristotle's argument is valid because while the supposition is not possible with respect to natural potencies it is possible *simpliciter* with respect to divine potency ... When it is said that Aristotle should not have accepted this supposition as possible because he did not believe that it is possible simply or through natural or supranatural potencies, I answer that Aristotle to a great extent agreed with our true faith and firmly believed that many things are impossible with respect to natural potencies and simply possible through supernatural potency. (Ibid.)

Buridan was right in arguing that his idea of counterfactual possibilities is quite different from the abstract possibilities in Averroes and Aquinas. To Buridan, it would seem strange to call something possible even while admitting that it cannot in any way be actualised. His criticism is not quite to the point, however. Aquinas and Averroes both remarked that in constructing Aristotle's indirect proofs, in which something impossible is deduced from the premises, one could formulate the argument as a hypothetical proposition and then deduce the denial of the antecedent from the denial of the consequent. There is no need to assume the actuality of what is impossible *per accidens*.[51]

In the current literature on late medieval philosophy, it has now become standard to state that the new modal theory developed in the early 14th century, with its emphasis on synchronic alternatives and the centrality of compossibility and co-assertability, was especially conducive to the devising of all kinds of thought experiments in natural philosophy. That this is indeed the case can be ascertained by looking at something like Nicholas Oresme's massive volume *On the Heavens and Earth*, which is replete with all kinds of cosmological thought experiments, all of them justified by way of reference to God's omnipotence and the notion that an alternative possible world could include features quite divergent from ours.[52] All the same, it is notable that an alternative model for conceptualising thought experiments existed in 12th and 13th century philosophy in the form of the abstractive approach favoured by Averroes and Aquinas, and that this model was developed to a great degree of sophistication and was capable of making sense of a wide range of thought experiments without reverting to the notion of actualisation in a possible world. We do not possess the means to ascertain conclusively, how Aristotle meant his indirect arguments to be read and understood; but considering that Buridan and Oresme took constant recourse in the omnipotence of God,

---

[51] Averroes, *Questions in Physics*, VII.2, 308ra; Aquinas, *In octo libros*, VII.1, 889.
[52] For remarks see Kukkonen, "The Impossible," 461–463.

which would have been a foreign concept to Aristotle, it is at least possible that Averroes' and Aquinas' reading is closer to Aristotle's way of thinking. Aristotle's scientific programme is, after all, concerned with mapping out the necessary and contingent properties of beings: and within such a programme, wondering about the possibilities open to the genus in relation to the species, and the species in relation to the individual, makes a certain degree of sense.

GALILEO'S USE OF MEDIEVAL THOUGHT EXPERIMENTS

Carla Rita Palmerino*

Galileo occupies a prominent place in the literature on thought experiments. The short argument that Salviati uses, in the *New Sciences*, to prove "without further experience" that Aristotle was wrong in assuming that a heavier body would fall more rapidly than a lighter one, has been the object of much scholarly interest. James Brown has singled it out as a perfect example of a Platonic thought experiment, which "gives a grip on nature just by thinking";[1] John Norton has used it to prove, against Brown, that thought experiments are just logical arguments in disguise;[2] Tamar Szabó Gendler, in turn, has argued that good thought experiments like Galileo's derive their persuasive power from the fact that they enable one to systematise "a previously inarticulable knowledge about the world"; whereas David Atkinson and Jeanne Peijnenburg have maintained that Galileo's argument is "logically deficient" and fails "to adequately describe the empirical world."[3]

That Galileo's thought experiment was inspired by a similar argument in Giambattista Benedetti's *Speculationum liber* (1585), is usually regarded as an insignificant historical detail. But if one looks at the many thought experiments presented in Galileo's work, one finds that many of them derive from pre-existing sources, notably Aristotelian ones. As I shall show, this fact must not be read as a sign of continuity between scholastic and early modern science, but rather as an indication that Galileo consciously used a traditional way of arguing in order to build up his non-traditional conclusions. I hence cannot share the opinion of James McAllister, who regards Galileo's thought experiments as a new

---

* Research for this article was made possible through the financial support of the NWO project "Visualizing the Invisible." I wish to thank Sophie Roux for a very fruitful discussion at the Grenoble conference, and Delphine Bellis, Christoph Lüthy, Lodi Nauta, Alexis Smets and Hans Thijssen for their comments on an early draft of this paper.

[1] Brown, "Thought Experiments," 528. Cf. also Brown, *Laboratory*; idem, "Platonic Account"; idem, "Thought Experiments Transcend."

[2] Norton, "Are Thought Experiments"; idem, "Why Thought Experiments."

[3] Gendler, "Galileo," 415; Atkinson and Peijnenburg, "Galileo and Prior," 127. Galileo's thought experiment is also discussed in the contribution to this volume by Goffi and Roux.

instrument, the validity of which was not recognised by contemporary Aristotelians:

> The discrepancy between what occurs in Galileo's thought experiments and what occurs under natural circumstances ensured that Aristotelian natural philosophers failed to attribute evidential significance to the former. For them, Galileo's thought experiments carried no weight: evidence in natural philosophy was constituted by natural occurrences.[4]

It is certainly not the case that Galileo's Aristotelian contemporaries relied exclusively on the evidence provided by natural occurrences. In the medieval period natural philosophy had turned into a "philosophy without nature" (in Murdoch's felicitous formulation), which heavily relied on arguments *secundum imaginationem*.[5] Following in Murdoch's footsteps, other scholars have emphasised the non-empirical character of medieval science; according to Peter King "thought experiments are the methodology of medieval science,"[6] whereas Edward Grant has observed that "the most powerful tool medieval natural philosophers possessed was not empiricism manifested by observation *per se*, but rather experience as adapted for use in thought experiments."[7]

As I shall show in the following pages, Galileo was acquainted—though possibly but indirectly—with the medieval *secundum imaginationem* arguments, and he consciously reused some of them with the aim of either strengthening or subverting the conclusions reached by their first proponents.

The medieval thought experiments that reappear in Galileo's works can be divided into two groups, namely those that concerned the actual physical world, and those that hinged on worlds that God, in his infinite potency, could have created. Arguments of the first type served Galileo to show that medieval natural philosophers had detected some major inconsistencies in Aristotle's physics, but had not understood the disruptive implications of their critique. Arguments of the second type were instead used to show that what medieval authors had imagined to happen in alternative worlds could actually be the case in our own physical world. Importantly, by reinterpreting old cosmological thought experiments in the light of his new science of motion, Galileo took an important step

---

4 McAllister, "Virtual Laboratory," 51. Cf. also McAllister, "Evidential Significance."

5 Murdoch, "The Analytic Character."

6 King, "Medieval Thought-Experiments."

7 Grant, "Medieval Natural Philosophy," 167. For an analysis of medieval *secundum imaginationem* arguments, cf. Grellard's contribution to this volume.

towards the unification of celestial and terrestrial physics. We shall also see that Galileo explicitly rejected thought experiments that made appeal to divine omnipotence. In his view, the only possible way to reason about imaginary scenarios was to explain them in conformity with the ordinary course of nature. For nothing meaningful could be said about a situation in which the laws of physics were supposed to have been violated.

1. *Reasoning about Our World: Galileo and Medieval Thought Experiments on Motion*

One part of Aristotle's physics that was questioned by medieval natural philosophers was his explanation of projectile motion. In *Physics* VIII, 10, Aristotle had asked the following question: "If everything that is in motion is being moved by something, how comes it that certain things, missiles for example, that are not self-moving nevertheless continue their motion without a break when no longer in contact with the agent that gave them motion?"[8] Aristotle came to the conclusion that projectiles move for some time after having lost contact with their projector because they are pushed by the air, which in turn receives its motion from the thrower. When this motion, which is communicated from one layer of air to the next, exhausts itself, the projectile falls to the ground.

Medieval authors such as Buridan and Oresme proposed an alternative explanation of projectile motion. They assumed that the thrower communicated to the projectile an *impetus*, which would be gradually consumed due to the resistance of the air and the internal gravity of the body. In his *Quaestiones super Physicorum libros*, John Buridan refuted Aristotle's explanation of projectile motion by means of three so-called "experiences," which concerned common objects such as a smith's mill, a lance, and a ship. While the experiences with the mill and with the ship look very much like everyday observations, that of the lance is presented as a "reasoned" experience.

> The second experience is this: A lance having a conical posterior as sharp as its anterior would be moved after projection just as swiftly as it would be without a sharp conical posterior. But surely the air following could not push a sharp end in this way, because the air would be easily divided by the sharpness.[9]

---

[8] Aristotle, *Physics* 266b28–30.

[9] John Buridan, *Quaestiones super Physicorum libros*, VIII, q. 2, quoted in Grant, *Source Book*, 275–276.

The "experience" is formulated in the conditional so as to suggest that it has not been executed. Buridan seems to assume that it is enough to have observed lances fly once to understand that the behaviour of one with a conical posterior cannot be that different from that of one with a flat posterior. In his book on thought experiments, Roy Sorensen lists Buridan's example of the lance under the category of unimprovables, that is to say, of those experiments that are not executed because of the absence of any evidential gain.[10] It is enough to form a mental picture of the situation under scrutiny to understand how things must stand.

In a section of his *De motu* devoted to explaining that air cannot act as a propelling force, Galileo invoked arguments similar to Buridan's. One of them brings into play an arrow that flies in the face of a strong wind; another one refers to a ship moving against the course of the current; and a further one involves a marble sphere, which can rotate for a long time, although "the air was not set in motion by the mover."[11]

Although the examples used by Buridan and Galileo are not exactly alike, the fact that they involve the same objects, namely an arrow, a ship, and a rotating wheel, suggests that they were part of a set of standard answers to Aristotle. Moreover, by using the expression "there is the argument of the arrow," Galileo indicated that he was consciously drawing on a set of standard objections.

The *De motu* is a juvenile work, still partly embedded in the Aristotelian tradition, and it therefore comes as no surprise that it contains some scholastic arguments that Galileo probably discussed with his colleagues at the University of Pisa. But it is interesting for us to see that a similar piece of reasoning reappears in the *Dialogue*. In the Second Day of the work, Galileo's spokesman, Salviati, tries to convince the Aristotelian Simplicio that "the medium impedes, rather than helps, the motion of projectiles." At a certain point Sagredo, the third interlocutor, interrupts the conversation and presents the following case:

> SAGR.: If two arrows were shot with the same bow, one in the usual way and one sideways—that is, putting the arrow lengthwise along the cord and shooting in that way—I should like to know which one would go the farther? ...
>
> SIMPL.: I have never seen an arrow shot sideways, but I think it would not go even one-twentieth the distance of one shot point first.[12]

---

10 Sorensen, *Thought Experiments*, 197.

11 *De Motu antiquiora* (1590 ca.), in Galilei, *On Motion*, 77–78.

12 Galilei, *Dialogue*, 152–153.

It is not an insignificant fact that the argument of the *De motu* reappears in the mouth of Sagredo, who often represents earlier stages of Galileo's own intellectual development. Simplicio's answer, which clearly reveals the imaginary character of the experiment, permits Sagredo to develop his argument in the desired direction:

> SAGR.: Since this is just what I thought, it gives me occasion to raise a question between Aristotle's dictum and experience. For as to experience, if I were to place arrows upon that table when a strong wind was blowing, one in the direction of the wind and the other across it, the wind would quickly carry away the latter and leave the former. Now apparently the same ought to happen with two shots from a bow, if Aristotle's doctrine were true, because the one going sideways would be spurred on by a great quantity of air moved by the bowstring—as much as the whole length of the arrow—whereas the other arrow would receive the impulse from only as much air as there is in the tiny circle of its thickness. I cannot imagine the cause of such a disparity, and I should like very much to know it.[13]

As Galileo often repeats in the *Dialogue*, the only way to reason about things one cannot directly observe is to explain them by analogy with familiar phenomena. This is why Sagredo provokes Simplicio into giving a reason for the fact that his answer does not correspond with what is observed in a situation which is analogous to the one under scrutiny.

> SIMPL.: The cause is obvious to me; it is because the arrow shot point foremost has to penetrate only a small quantity of air, and the other has to cleave as much as its whole length.
>
> SAGR.: Oh, so when arrows are shot they have to penetrate the air? If the air goes with them, or rather if it is the very thing which conducts them, what penetration can there be? Do you not see that in such a manner the arrow would be moving faster than the air? Now what conferred this greater velocity upon the arrow? Do you mean it to say that air gives it a greater speed than its own? ...
>
> SALV.: How many propositions I have noted in Aristotle that are not only wrong, but wrong in such a way that their diametrically opposites are true, as happens in this instance![14]

Contrary to his medieval predecessors, Galileo does not limit himself to showing that the imagined behaviour of the arrow contradicts Aristotle's theory of projectile motion. By letting Simplicio try to defend Aristotle's view, he sheds light on the main inconsistency of the latter, namely the

---

[13] Ibid., 153.
[14] Ibid., 153–154.

fact that it treats the medium both as a propelling force and a resisting factor. Galileo is clearly interested in stressing the counterintuitive character of Aristotle's hypothesis, which in this particular case is not in conformity with, but rather runs counter to, natural occurrences. Incidentally, this is not the only passage in Galileo's work where Aristotle's theory of motion is criticised for being in contradiction with everyday experience. In the First Day of the *Two New Sciences* we read, for example:

> SALV.: Aristotle says: 'A hundred pound iron ball falling from the height of a hundred *braccia* hits the ground before one of just one pound has descended a single *braccio*.' I say that they arrive at the same time. You find, on making the experiment, that the larger anticipates the smaller by two inches; that is, when the larger one strikes the ground, the other is two inches behind it. And now you want to hide, behind those two inches, the ninety-nine *braccia* of Aristotle, and speaking only of my tiny error, remain silent about his enormous one.[15]

In this passage, Salviati stresses the fact that even though Galileo's law holds fully true only for bodies falling in an imaginary void space, it is still more in agreement with what is seen happening in physical reality than Aristotle's theory, which aimed at describing the "real behaviour" of bodies falling in the air. Simplicio tries to defend Aristotle by stating that "perhaps from very great heights, of thousands of *braccia*, that would follow which is not seen at these lesser heights." To this, Salviati replies:

> If that is what Aristotle meant, you saddle him with a further error that would be a lie. For no such vertical heights are found on earth, so it is clear that Aristotle could not have made the trial; yet you want to persuade us that he did so because he says that the effect 'is seen.'[16]

This is only one of the many passages in which Galileo stresses the hypocrisy of the Aristotelians' appeal to the heuristic power of sense perception; the experiences invoked by the scholastics are, more often than not, pure mental fictions.

We have seen that Galileo uses the argument of the arrow not simply to correct Aristotle's theory of motion, but to subvert it. Indeed, Sagredo points out to Simplicio that

> this whole thing takes place just exactly the opposite to what Aristotle says, and that it is as false that the motion confers motion upon the projectile as it is true that it is this *alone* which impedes it.[17]

[15] Galilei, *Two New Sciences*, 69.
[16] Ibid.
[17] Galilei, *Dialogue*, 153.

Although Galileo interprets the thought experiment in the same manner as his medieval predecessors, the conclusions he draws from it are much more radical. The partisans of the impetus theory thought, contrary to Aristotle, that projectiles did not require the application of an external force to continue to move, but they did not dispute the fact that their motion, being violent, must come to an end. Galileo reasons instead as follows: if it is true that the air is not a propelling force, but a resisting factor, shouldn't we conclude that a projectile thrown in the void would continue to move endlessly? In Galileo's new physics motion is not a process, which is per definition transitory, but a state which would only come to an end due to the action of external impediments.

The difference between the Galilean and the medieval conception of motion becomes even clearer if we look at another thought experiment, which contrary to the previous one involves a physically impossible situation. What would happen if a stone fell through a tunnel traversing the centre of the earth? Would it come to rest? And if so, where?

During the Middle Ages some natural philosophers argued, in accordance with Aristotle's theory, that the stone would stop as soon as it reached its natural place, namely the centre of the universe, which coincided with the centre of the earth. But partisans of the impetus theory, such as Albert of Saxony and Nicholas Oresme, argued that

> if an opening were made from here to the centre of the earth and beyond and a heavy object fell through this opening or hole, upon reaching the centre it would pass beyond and begin to go upward by means of this accidental and acquired property [the *impetus*]; then it would fall back again and come and go several times just as we can observe in the case of a heavy object hanging from a beam by a long chord.[18]

> And so it would be moved oscillating about the centre until there no longer would be any such impetus in it, and then it would come to rest.[19]

The case of the heavy body falling down an imaginary tunnel passing through the centre of the earth was discussed by a number of 16th century authors, such as Erasmus of Rotterdam, Francesco Maurolico, Alessandro Piccolomini and Francesco Buonamici.[20] The latter, who was professor at Pisa during Galileo's student days, invoked the thought

---

[18] Oresme, *Le livre du ciel*, 145. This passage is discussed in Clagett, *Science of Mechanics*, 570.

[19] Albert of Saxony, *Questiones in libros de coelo et mundo Aristotelis*, quoted in Clagett, *Science of Mechanics*, 566.

[20] See Besomi and Camerota's account in Galilei, *Dialogo sopra i due massimi*, 2:178.

experiment in chapter 44 of the second book of his monumental *De motu libri X*, which is devoted to proving that the upward and downward motions of a heavy body derive from contrary principles. Buonamici compares the behaviour of the stone falling through the imaginary tunnel to that of a ball rolling first down and then up a V-shaped plane. Just like the ball must rest for a moment at the lowest point of the plane, so the body falling in the tunnel must also pause at the centre of the earth, for this is the point in which its motion turns from natural into violent.[21]

The thought experiment of the ball falling through a perforated earth is discussed more than once in Galileo's *Dialogue*. We encounter it for the first time at the beginning of Day One, when Sagredo uses it to support Salviati's idea that "the impetus acquired by a falling body at any point of its motion is enough to carry it back to the height from which it started."[22] Sagredo bases his analysis of this imaginary case on the observation of two more familiar phenomena:

> SAGREDO: If the earth were tunnelled through the centre, and the ball were let fall a hundred or a thousand yards toward the centre, I verily believe that it would pass beyond the centre and ascend as much as it had descended. This is shown plainly in the experiment of a plummet hanging from a cord, which, removed from the perpendicular (its state of rest) and then set free, falls toward the perpendicular and goes the same distance beyond it—or only so much less as the cord, the resistance of the air, and other accidents impede it. The same thing is shown by water which, descending through a siphon, climbs back up as much as it went down.[23]

Like Nicholas Oresme and Albert of Saxony, Galileo compares the fall of the stone through the tunnel to the oscillation of a pendulum. The cogency of the analogy is guaranteed by the fact that, although the stone in its imaginary fall keeps following a perpendicular line, its motion turns from one of descent into one of ascent, once it surpasses the centre of the earth. But Sagredo's argument differs in one important point from that of the partisans of the *impetus* theory. While the latter assume that the stone would come to a stand still after a number of oscillations, Galileo believes that it would perpetually move to and fro, as a pendulum does in the absence of impediments. But of course, the motion of a pendulum in the absence of impediments is in turn something that is never directly observed.

---

[21] Buonamici, *De motu*, 220–223.
[22] Galilei, *Dialogue*, 22.
[23] Ibid., 22–23.

The thought experiment is invoked once again in Day Two of the *Dialogue*, where it is used to argue that the "principle which moves heavy projectiles upward ... is no less internal and natural than that which moves them downward."[24] To Simplicio, who claims that heavy bodies are drawn by their internal principle towards "the centre of the universe and of the earth," Salviati asks to imagine that "the terrestrial globe were pierced by a hole which passed through its centre" and a cannon ball were dropped into this hole.

> SALVIATI: Having arrived at the centre is it your belief that it would pass on beyond, or that it would immediately stop its motion there?
>
> SIMPLICIO: I think it would keep on going a long way.
>
> SALV.: Now wouldn't this motion beyond the centre be upward, and according to what you have said preternatural and constrained? But upon what other principle will you make it depend, other than that which you have already called intrinsic and natural? Let me see you find an external thrower who shall overtake it once more to throw it upward. And what is said thus about motion through the centre is also to be seen up here by us. For the internal impetus of a heavy body falling along an inclined plane which is bent at the bottom and deflected upward will carry the body upward also, without interrupting its motion at all. A ball of lead hanging from a thread and moved from the perpendicular descends spontaneously, drawn by its internal tendency; without pausing to rest it goes past the lowest point and without any supervening mover it moves upward. I know you will not deny that the principle which moves heavy bodies downward is as natural and internal to these which moves light ones upward is to those.[25]

Giovanni Barenghi, author of the *Considerazioni sopra il Dialogo de due massimi sistemi Tolemaico, e Copernicano* (Pisa, 1638), a book written in defence of the principles of Aristotle's cosmology, expressed his astonishment at the fact that Galileo "who had copied all these difficulties from chapters 44 and 45 of the second book of Buonamici's *De motu*, had not wanted to see the solutions which are given in the same places."[26] The problems listed by Salviati had in fact already been formulated by Buonamici, who, as we have seen, had tried to solve them by asserting the existence of a moment of rest separating the upward and the downward motion. But such an answer must have appeared to Galileo

[24] Ibid., 235.
[25] Ibid., 236.
[26] Barenghi, *Considerazioni*, 121.

so clearly *ad hoc* that he did not even find it necessary to mention it in his *Dialogue*:

> SIMPLICIO: I believe there are answers for all these objections, though for the moment I do not remember them.[27]

2. *Reasoning about Alternative Worlds: Galileo and the Scholastic Arguments* de potentia Dei absoluta

As mentioned in the introduction, medieval thought experiments often involved scenarios which were regarded as impossible in the ordinary course of things, but not from the point of view of God's *potentia absoluta*. In his *Theology and the scientific imagination from the Middle Ages to the seventeenth century*, Amos Funkenstein explains well how medieval arguments *secundum imaginationem* cleared the way for future conceptual developments:

> We ought, then, to pay close attention to the terms in which a theory defines 'improbabilities' and, still more important, 'impossibilities.' The more precise the argument, the likelier it is to be a candidate for future revisions. Once the impermissible assumption is spelled out with some of its consequences, it is but a matter of time and circumstances ... until the truly radical alternative is reconsidered ... In effect, many of the "absurdities" in terms of Aristotelian science had already become, through the medieval exercise in hypothetical reasoning, ... possibilities *de potentia Dei absoluta*, only incompossible with our universe. With the usage of ideal experiments in the seventeenth century many such incompossibles became limiting cases of our universe.[28]

In the following I shall explain, by means of a number of examples, how what was an absurdity according to Aristotle's physics could, in late medieval natural philosophy, become a possibility *de potentia Dei absoluta*, to be even later transformed by Galileo into a physical possibility.

I shall again take Galileo's *Dialogue* as a starting point. In a passage of the First Day, Salviati wonders whether the fact that the moon, the sun, and other celestial bodies have a round shape might not be a sign that their parts, like those of the Earth, have a natural tendency to come together.

---

[27] Galilei, *Dialogue*, 237.
[28] Funkenstein, *Theology*, 16–17.

> If at any time one of these parts were forcibly separated from the whole, is it not reasonable to believe that it would return spontaneously and by natural tendency? And in this manner we should conclude that straight motion is equally suitable to all world bodies.[29]

To this, the Aristotelian Simplicio reacts by pointing out, first, that the reason why on the earth heavy objects fall downwards is not because they want to reunite with the whole, but because they are directed toward the centre of the universe, which by accident coincides with the centre of the earth. As for the imaginary scenario invoked by Salviati, Simplicio thinks that

> it is vain to inquire ... what a part of the globe of the sun or moon would do if separated from its whole, because what you inquire into would be the consequences of an impossibility. For, as Aristotle also demonstrates, celestial bodies are invariant, impenetrable, and unbreakable; hence such a case could never arise. And even if it should, and the separated part did return to its whole, it would not return thus because of being heavy or light, since Aristotle also proves that celestial bodies are neither heavy nor light.[30]

Although in these lines Simplicio seems to reject off hands the possibility of analysing states of affairs, which cannot be the case in our world, it was a customary habit of scholastic philosophers to reason about counterfactual situations. Aristotle himself invoked impossible scenarios, although this happened mostly in *reductio ad absurdum* arguments. So, for example, in *Physics* IV, 8, he deduces the impossibility of the void from the consideration of how bodies would move in an imaginary empty space; in *De caelo* I, 8–9, he excludes the existence of a plurality of worlds by reasoning that if these worlds existed they could not be different from one another and hence they would all occupy the same place; similarly, in *De caelo* II, 14, he argues that if the earth rotated, its movement would have to be continually enforced, given that the natural motion of its parts is not circular, but rectilinear, "and therefore being enforced and unnatural, it could not be eternal."[31]

When they are not self-refuting, Aristotle's thought experiments seem to beg the question. So in *De caelo* IV, 3 he tries to prove that the motion of heavy bodies is directed towards the centre of universe, by arguing that "if the earth were removed to where the moon is now, separate parts

---

[29] Galilei, *Dialogue*, 33–34.

[30] Ibid., 34.

[31] Aristotle, *De caelo* 296a24–34. Cf. Knuuttila and Kukkonen's essay in this volume, which deals with ancient thought experiments starting with impossible premises.

of it would not move towards the whole, but towards the place where the whole is now" (that is to say they would continue falling towards the centre of the universe).[32] And again, in order to demonstrate that the centre of the earth coincides *per accidens* with the centre of the universe, he argues that if a weight many times its own were added to one hemisphere of the earth, the latter would move "until it possesses the centre with its own centre, for its impulse extends to that point."[33]

Especially after the condemnation of 1277, medieval natural philosophers exerted their imaginative power in reasoning about scenarios that were incompatible with the principles of Aristotle's physics, but possible from the point of view of the *potentia Dei absoluta*. Article 147 condemned the following proposition: "That what is simply impossible cannot be done by God or by another agent," arguing that "it is an error, if impossible is understood according to nature." The existence of a void space or of a plurality of worlds, which Aristotle had both rejected, clearly did not imply a logical contradiction. As a consequence, 14th century authors did not endorse Aristotle's *reductio ad absurdum* arguments, but granted to God the power of creating many worlds or a space devoid of matter. At the same time, however, they tried to preserve the validity of Aristotle's physics for our actual world, by subscribing to his theory of elements, by reasserting the coincidence between centre of the universe and centre of the world, and by denying that a void space could be created under natural circumstances. As we shall see in the following pages, Galileo's strategy was to show that the medieval arguments *de potentia Dei absoluta* could be used to falsify precisely those tenets of Aristotle's physics which 14th century natural philosophers were not willing to relinquish.

2.1. *Does the Universe Have More Than One Centre?*

All 14th century natural philosophers who admitted the possibility of a plurality of worlds felt the need to rebut Aristotle's argument that these worlds would converge to the same place. Buridan, for example, devoted one of his *Quaestiones in Aristotelis De caelo* to establishing "whether, if there existed several worlds, the earth of one world would move naturally towards the earth of the other." He answered the question in the negative, arguing that each world would have its own order "established by God by

[32] Ibid., 310b1–5.
[33] Ibid., 297b5–8.

means of the heaven" (*a Deo mediante caelo*), so that each earth would tend towards the centre of its own world, and by no means towards the centre of our world.[34] Similarly, Albert of Saxony reasoned that if God, in his absolute power, created many worlds,

> just as the earth is naturally at rest at the centre of our world, so any other earth would be at rest at the centre of its own world, ... and if it were elongated from it, it would tend towards it, because this would be the nearest centre. But if it were equidistant between two centres, it would remain at rest, like a piece of iron between two equally attracting magnets.[35]

A similar argument is put forward by Oresme in Book I, chapter 5, of his *Livre du ciel et du monde*, where we read that

> If a portion of earth were equidistant between two worlds and if it could be separated, one part would go to the centre of one world and the other portion to the centre of the other world. If the portion could not be divided, it would not move at all because of the lack of inclination, being like a piece of iron halfway between two magnets of equal strength.[36]

In his *Livre du ciel et du monde*, Oresme shows greater awareness than either Buridan or Albert of Saxony concerning the implications that the hypothesis of a plurality of worlds has for Aristotle's doctrine of natural places. In book I, chapter 24, he explains that "if God created another world like our own, the earth and the other elements of this other world would be present there just as they are in our own world." This means that heavy bodies "would form a single mass possessed of a single place and would be arranged in up and down order as we have indicated, just like the mass of heavy bodies in this world."[37] But what is more, Oresme seems also to deny the existence of absolute directions within our world. As he explains, the terms *up* and *down* "indicate nothing more than the natural law concerning heavy and light objects, which is that all the heavy bodies *so far as possible*, are located in the middle of the light bodies."[38] In order to prove that elements do not possess

---

[34] Buridan, *Expositio*, 329.

[35] "Sicut terra in isto mundo quiescit in medio huius mundi naturaliter, ita terra alterius mundi quiesceret in medio alterius mundi naturaliter, et sic una non tenderet ad aliam, sed ad medium sui mundi, si essent extra ipsum, quia tenderet ad medium sibi propinquius. Sed si contingeret quod aequaliter distare ab utroque, quiesceret inter ambo, sicut ferrum inter duos adamantes aequaliter trahentes." Albert of Saxony, *Quaestiones*, 142.

[36] Oresme, *Le livre du ciel*, 135.

[37] Ibid., 174–175.

[38] Ibid., 173.

"within themselves the two motive qualities of weight and lightness in the absolute manner that, in the opinion of some people, two contraries have in relation to their intermediate state," Oresme makes use of an interesting thought experiment.[39] He argues that if a pipe extending from the centre of the earth to the heavens were filled with fire except for a very small amount of air at the top, the air would descend; if instead the pipe were filled with water except for some air near the centre of the earth, the air would mount up to the heavens.[40] This example, to which Oresme refers again and again in his book,[41] shows that one and the same element can have different natural motions in different media, which means, in turn, that elements, considered in isolation, cannot be regarded as either heavy or light.[42] As John North has noted, this statement must be regarded as a "small but useful contribution to the slow erosion of the doctrine of natural place, and hence directly to that of a distinction between terrestrial and celestial sciences."[43]

That Oresme's theory of elements has important cosmological implications becomes clear in book II, chapter 25, which presents "the opinions of some thinkers about the movement of the earth." Here, Oresme first argues, by means of arguments that bear a striking resemblance to those used by Galileo in the *Dialogue*, that it is impossible to demonstrate "by any experience" that the daily rotation belongs to the heavens and not to the earth.[44] Then, he moves on to rebut Aristotle's argument according to which "circular motion is not natural to the earth, for it has another motion," namely the "rectilinear motion downward" of its parts.[45]

> As for the argument where it is stated that every single body has a single simple motion ... we can cite the case of air which moves downward when it is in the region of fire and upward when it is in the region of water, both being simple movements. Therefore, we can say with an even greater

[39] Oresme, *Le livre du ciel*, 687. A similar argument was phrased by Ockham, who maintained that if God created many worlds, the behaviour of their respective elements would be analogous to that of two fires, the one moving over Oxford, the other one over Paris. If the masses of fire were switched, they would have no tendency to move back to their former places. Cf. Grant, "Condemnation of 1277," 221.

[40] Oresme, *Le livre du ciel*, 71.

[41] Ibid., 697; cf. also 71, 173, 529.

[42] "Therefore, I say that a heavy body to which no light body is attached would not move of itself, for in such a place as that in which this heavy body is resting, there would be neither up nor down because, in this case, the natural law stated above would not operate." (Ibid., 173).

[43] North, *Stars, Minds, and Fate*, 302.

[44] Oresme, *Le livre du ciel*, 529–527.

[45] Ibid., 528.

> show of reason that each simple body or element of the universe, with the possible exception of the sovereign heaven, moves in its proper place with circular motion. If any part of such a body is out of its place or outside the main body, it returns to it as directly as it can, once the hindrance is removed; this would surely happen if some part of the heavens were to get outside. It is not inconvenient that a simple body has its own simple motion in its proper place and another motion in its parts when they return to their proper place.[46]

The similarity of this argument to that proposed by Salviati and rejected by Simplicio in the *Dialogue* is noteworthy. Like Galileo 250 years later, Oresme not only maintains that the natural motion of a body can be different from the natural motion of its parts, but also seems to imply that the reason why the parts of the earth cohere is the same as that for which the parts of other celestial bodies cohere.

On the basis of the passages analysed so far it would seem as if there were sufficient grounds to endorse Duhem's famous judgment about the revolutionary character of Oresme's natural philosophy:

> The earth's weight does not require it to remain stationary at the centre of the universe, as it does for Aristotle's physics; surrounded by elements, the lighter ones containing the heavier ones, it is free to move about space in the manner of a planet. In any case, nothing prevents each planet being formed out of a heavy earth surrounded by water, air, and fire analogous to ours. This new theory allows one to compare the earth and the planets while the Peripatetic theory absolutely prohibited it.[47]

Although his theory of elements would have allowed him to displace the earth from the centre of the universe, Oresme did however not take this step. The possibility that the earth might revolve around the sun is not even contemplated in the *Livre du ciel et du monde*, and also the hypothesis of the daily rotation is in the end discarded as being "more against natural reason than all or many of the articles of our faith."[48] What is more, in the second half of his work, Oresme aligns himself with the principles of Aristotelian cosmology: he states explicitly that the centre of the earth coincides with the centre of the universe and supports this view by means of two thought experiments borrowed from the *De caelo*. In book II, chapter 30, he maintains that if a small quantity of matter were added to the earth, the latter "would be moved slightly until the centre of

---

[46] Ibid., 528–529.
[47] Duhem, *Medieval Cosmology*, 477.
[48] Ibid., 539.

its weight reached the centre of the world,"[49] and in book IV, chapter 3, he expresses the conviction that if we were to transport the earth to the place occupied by the moon, leaving only a small part at its place, "this small part would not be moved toward the earth occupying the moon's place, but rather the earth would be moved back to its proper location."[50]

The question of what would happen if the earth were removed from its place is discussed by Galileo in the *Dialogue*. Salviati first summarises Simplicio's position according to which

> just as parts of the earth, removed from the whole ... return to their places naturally and spontaneously in a straight motion, so it may be inferred (granted that *eadem sit ratio totius et partium*) that if the terrestrial globe were forcibly removed from the place assigned to it by nature, it would return by a straight line,[51]

and then comments that

> if it should be said that the parts of the earth do not move so as to go toward the centre of the universe, but so as to unite with the whole earth (and that consequently they have a natural tendency toward the centre of the terrestrial globe, by which tendency they cooperate to form and preserve it), then what other 'whole' and what other 'centre' would you find for the universe, to which the entire terrestrial globe would seek to return if removed therefrom, so that the rationale of the whole might still be like that of its parts? I might add that neither Aristotle nor you can ever prove that the earth is *de facto* the centre of the universe ... [52]

As Salviati rightly points out, Simplicio's argument presupposes what it wants to prove, namely that the earth is located at the centre of the universe. But what if the reason why bodies fall downward was instead that they want to reunite with their whole?

> SALV.: Now just as all the parts of the earth mutually cooperate to form its whole ... why may we not believe that the sun, moon and other world bodies are also round in shape merely by a concordant instinct and natural tendencies of all their component parts? If at any time one of these parts were forcibly separated from the whole, is it not reasonable to believe that it would return spontaneously and by natural tendency?

While Simplicio, following Aristotle, reasons from part to whole, that is to say from the behaviour of a piece of earth to the behaviour of the whole earth, Salviati reasons from part to part, that is to say, he infers

---

49 Ibid., 561.
50 Ibid., 677.
51 Galilei, *Dialogue*, 33.
52 Ibid.

the behaviour of a part of a planet from the behaviour of a part of the earth. As we have seen, Oresme had already maintained that a piece of a celestial body, if displaced, would return to its whole by the shortest possible line (i.e., by falling down rectilinearly), but had not developed his reasoning to its logical consequences. The same argument is used in the *Dialogue* to assert the existence of many centres of gravity within our world and at the same time to reject the Aristotelian claim—which was instead endorsed by Oresme—that falling bodies are directed towards the centre of the universe.

2.2. *Galileo's Critique of the Arguments* de potentia Dei absoluta

Galileo's Aristotelian contemporaries also invoked *secundum imaginationem* arguments in order to prove that the earth was located at the centre of the universe, which was the place towards which heavy bodies were directed. Buonamici, for example, argued in his *De motu* that if the earth were smashed up and dispersed, its parts would precipitate towards the centre of the universe and would heap up like the stones of a vault.[53] Similarly, the German Jesuit Johann Georg Locher, in a dissertation written under Christoph Scheiner's supervision, argued that, if it were true as Copernicus had maintained, that the parts of the earth naturally moved with circular motion and that they fell downward just in order to reunite with their whole, then

> if the whole earth, together with the water, were reduced to nothing, no hail or rain would fall from the clouds, but would only be carried naturally around; nor would any fire or flaming thing ascend, since in their view, probably, there is no fire above. Which however, experience and reason refute.[54]

This passage, which Simplicio reads aloud in the Second Day of the *Dialogue* together with other excerpts from Locher's *Disquisitiones*, incites Salviati to the following ironic comment:

> SALV.: The providence of this philosopher is admirable and worthy of great praise, he not contenting himself with thinking of things that might happen in the course of nature, but trying to provide himself against occasions on which things happened which are absolutely known never to happen. Now I ought to give up, since he has such a big advantage over me; namely experience which I lack. For as yet I have never happened to

[53] See Galilei, *Dialogo sopra i due massimi*, 2:581.
[54] Galilei, *Dialogue*, 243.

> see the terrestrial globe and all the element of water annihilated, so as to have been able to see what the hail and water did in this little cataclysm.[55]

Salviati's critique of Locher's thought experiment is in a way comparable to the objections which have recently been raised against the analytical philosophers' habit of appealing to outlandish scenarios. In commenting a thought experiment by Hilary Putnam, which supposes that all domestic cats are in fact robots controlled by a clever deceiver, Nicholas Rescher has recently observed that "in the ordinary run of things, biology and behaviour run together." By "tearing apart what experience ... binds together," Putnam does not "clarify" a concept, but rather "abrogates it." By so doing, he creates a scenario in which "all bets are off."[56]

Galileo similarly regards Locher's thought experiment as one in which "all bets are off." Not only does the Jesuit tear apart things which one always experiences as connected (namely the earth and its parts), but he also gives too imprecise a description of his imagined scenario.

> SALV.: I should gladly pay to have a chat with this fellow, in order to ask him whether when this globe vanished it took away also the common centre of gravity, as I suppose it would. In that case, I think that the hail and water would remain senseless and stupid among the clouds, not knowing what to do with themselves. Or it might be that, attracted by such a large empty space left by the departure of the terrestrial globe, all the surrounding things would be rarefied ... and would rush with great speed to refill it. ... Or it might be that the same materials, being insensitive to light, would not discover the earth's departure, and would blindly descend as usual, expecting to encounter it. ... And finally, to give this philosopher a less indefinite reply, I say to him that I know just as much about what would happen after the earth was annihilated as he would have known about what was going to take place on it and around it before it was created. Since I am sure that he would say that he would not even have been able to imagine anything of what was to follow, since only experience has given him knowledge of this, he should not refuse to pardon me, but should forgive me for not knowing as much as he does about the things which would happen after the annihilation of this globe, seeing that I lack this experience which he possesses.[57]

The criterion, which Galileo applies in deciding about the feasibility of a thought experiment, is whether experience can serve as a guide in evalu-

[55] Ibid., 244.

[56] Rescher, *What if?*, 152–153.

[57] Galilei, *Dialogue*, 244–245. A similar remark is made with respect to another thought experiment proposed by Locher (cf. ibid., 245–246).

ating the imagined scenario. In his view, one can make a prediction about the behaviour of a stone falling through a perforated earth, because this state of affairs, albeit counterfactual, is physically analogous to situations we know from our ordinary experience. But the case is different with Locher's thought experiment, which pretends to explain what happens "in the ordinary course of nature" by appealing to a scenario in which the laws of nature are being subverted. So, when Simplicio reads another passage of the *Disquisitiones*, in which Locher supposes that the earth may be made to rotate by an angel or by God's continuous intervention, Salviati observes, with a tad of irritation, that "whatever begins with a Divine miracle or an angelic operation ... is not unlikely to do everything else by means of the same principle."[58]

2.3. *Does Nature Admit the Void?*

As mentioned before, another question on which 14th century natural philosophers departed from Aristotle's physics concerned the possibility of a void space. In this case, too, the influence of the Condemnation of 1277 should not be underestimated. The possibility of an extra-cosmic void was implied by Article 34 of the condemnation, which granted God the power of creating many worlds kept apart by a void space, and by Article 39, which prescribed that God could move the world rectilinearly leaving behind an empty space.[59] The possible existence of a void within the world was instead implied by Articles 139–141, which stated that God could make accidents exist without a subject or several dimensions exist simultaneously in the same place. Although these articles were aimed at safeguarding the dogma of transubstantiation, they also affected theories of place and motion. For one could argue, as Walter Burley did, that just as in the miracle of the Eucharist, the accidents of bread and wine subsisted without their substance, in similar manner a separate space could exist independently of a material body.[60]

In the wake of the condemnation, a number of *secundum imaginationem* arguments were proposed in which God was supposed either to annihilate or to displace the world, or to create a simple volume without there being a substance coexisting with it.[61] But what did 14th century

---

[58] Ibid., 237.

[59] Grant, *Much Ado about Nothing*, 108–110; idem, "Condemnation of 1277," 217–233.

[60] Grant, "Condemnation of 1277," 233.

[61] Ibid., 228–229; Duhem, *Medieval Cosmology*, 392–414.

natural philosophers have to say, then, about the many physical arguments against the void put forward by Aristotle?

In his *Quaestiones in octo libros Physicorum*, book IV, questions 11 and 12, Albert of Saxony asked himself "whether if a vacuum did exist, a heavy body could move in it" and whether it would move "with a finite velocity or local motion."[62] Albert answered both questions in the affirmative, arguing that Aristotle's argument that the motion in a void would be instantaneous held true for a simple body, but not for a "mixed or compound body," which had an internal resistance to motion and hence would move "successively."[63] As for the absence of an external resistance, Albert observed, in question 11, that:

> a heavy body is moved through air more quickly than through water because air is subtle and resists less than water, and since in a vacuum there would be absolutely no resistance, it seems that a heavy body ought to be moved most easily (*optime*) in a vacuum.[64]

In question 12, Albert reaches the interesting conclusion that mixed bodies of homogeneous composition but different weight "are moved with equal velocity in a vacuum but not in a plenum." The reason he adduces for this is that in the void "the ratio of motive power to total resistance in one body is the same as in another homogeneous body, because they both have only internal resistance," whereas in a plenum this ratio would be greater for the heavier body, which would hence descend faster.[65]

When reading Albert's analysis of the motion of bodies in different media, one is reminded of a famous thought experiment which Galileo presents in the First Day of his *Two New Sciences*:

> SALV.: We have seen that the difference of speed in moveables of different heaviness is found to be much greater in more resistant mediums ... This seen I came to the opinion that if one were to remove entirely the resistance of the medium, all materials would descend with equal speed.
>
> SIMPLICIO: That's a fine thing to say, Salviati. I shall never believe that even in the void—if indeed motion could take place there—a lock of wool would be moved as fast as a piece of lead.
>
> SALV.: ... We are trying to investigate what would happen to moveables very diverse in weight, in a medium quite devoid of resistance, so that the

[62] Grant, *Source Book*, 335, 338.
[63] Ibid., 335, 341.
[64] Ibid., 335.
[65] Ibid., 341.

> whole difference of speed existing between these moveables would have to be referred to inequality of weight alone. Hence just one space entirely void of air ... would be suitable for showing us sensibly that which we seek. Since we lack such a space, let us observe what happens in the thinnest and least resistant mediums, comparing this with what happens in others less thin and more resistant. If we find in fact that moveables of different weight differ less and less in speed as they are situated in more and more yielding mediums; and that finally, despite extreme difference in weight, their diversity of speed in the most tenuous medium of all (though not void) is found to be very small and almost unobservable, then it seems to me that we may believe, by a highly probable guess, that in the void all speeds would be entirely equal.[66]

Like Albert of Saxony, Galileo extrapolates the behaviour of bodies in a void from their observed behaviour in media of gradually decreasing densities, and like him he seems to believe that precisely in the absence of external resistance motion takes place most naturally. One obvious difference between Albert's and Galileo's analyses is that the former only applies to homogeneous bodies, whereas the latter extends to objects of different material. But another, more significant point of disagreement concerns the physical possibility of a void. In question 8 of book IV Albert explicitly denies the possibility of a natural vacuum, and in question 12, after affirming the possibility of a motion in a hypothetical void, he states that "if God should make a vacuum, natural beings would then strive infinitely to avoid it" so that "before the sky would allow a vacuum to remain, it would descend and fill it."[67]

Albert's position was by no means eccentric. Although they were ready to admit that God could create a void space and have objects move in it, 14th century natural philosophers remained faithful to the Aristotelian principle that nature abhors a vacuum, and denied the possibility that a vacuum, intended as an empty continuous space, could naturally exist "below the heavens." In order to support their view, they invoked a number of real and imaginary experiences. One of them referred to two marble or glass plates that are brought in contact with one another.

In his *De rerum natura*, Lucretius had argued that if two flat, smooth slabs lying on top of one another were separated, the void that was created between them would be filled by the air, but not instantaneously.

[66] Galilei, *Two New Sciences*, 75–76.
[67] Grant, *Source Book*, 339.

"For the air must occupy each point of space in succession before the whole is occupied."[68]

Lucretius' view was of course at odds with Aristotle's, who denied both that a void could be created in nature, and that a motion in the void could take place with a finite speed. The case of the separation of two plane surfaces was often discussed by medieval authors, who explained it in the most different ways, although they all tried to adhere to the principles of Aristotle's physics.

According to Roger Bacon, two perfectly polished plates that were in contact with one another could not be separated unless one managed to bend one of them into a curve.[69] Blasius of Parma stated instead that the two plates could be disjoined while remaining parallel to each other, because the successive separation would be accompanied by a concomitant insertion of air between them: "there was no first air that was first in the centre just as there is no first distance by which these plane bodies are now separated."[70] Albertus Magnus thought that upon separation air would instantaneously rush in to fill the space, whereas Walter Burley believed that a thin layer of air would always be present between the two plates and that the separation would be accompanied by a concomitant rarefaction of the air.[71]

As Edward Grant has rightly observed, it is irrelevant whether or not medieval authors tested the case, given that their theories were immune from verification. Roger Bacon, for example, would have interpreted all instances of separation as cases in which the surfaces were curved, and all instances in which the separation failed as cases in which they were not. "In the absence of any proper means of determining whether any two surfaces were truly inclined or not, Bacon used his theory to make the determination after the fact."[72]

In the *Two New Sciences*, Salviati presents the case of the two slabs as one that could easily be subjected to empirical verification:

> SALV.: We may see whenever we wish that two slabs of marble, metal, or glass, exquisitely smoothed, cleaned, and polished and placed one on the other, move effortlessly by sliding, a sure argument that nothing gluey joins them. But if we want to separate them while keeping them parallel, we meet

[68] Lucretius, *De rerum natura* I 385–389.

[69] Cf. Bacon, *Opera*, fasc. 13: *Physics* 226, quoted in Grant, *Much Ado about Nothing*, 87.

[70] *Quaestio de tactu corporum durorum*, quoted in ibid., 91.

[71] Ibid.

[72] Ibid., 88.

> with resistance; for the upper slab in being raised draws the other with it, and holds it permanently even if it is large and heavy. This clearly shows nature's horror at being forced to allow, even for a brief time, the void space that must exist between the slabs before the running together of parts of the surrounding air shall occupy and fill the space.[73]

The words Galileo puts in the mouth of Salviati could easily be subscribed to by an Aristotelian. Both would agree on the fact that the slabs offer resistance to separation and that this resistance is due to nature's fear of the void. In order to mark his distance from the Aristotelians, Galileo uses a purely rational argument, which he puts in the mouth of Sagredo. The latter argues that if air were able to fill instantaneously a void space,

> the two surfaces would be separated without any resistance whatever, the same instant of time sufficing for their separation and for the running together of the surrounding air to fill the void that might otherwise remain between them. Thus from the following of the upper slab by the lower, it is deduced that motion in a void would not be instantaneous. It is then further deduced that some void indeed does remain between the surfaces, at least for a very brief time; that is for as long as the time consumed by the ambient air in running to fill this void. For if no void existed there, neither would there be any need on the part of the ambient air of running together, or of any other motion.[74]

By putting his argument partly in Salviati's and partly in Sagredo's words, Galileo manages to keep the empirically ascertained fact distinct from its rational analysis. Sagredo's response to the Aristotelians is not based on new empirical evidence, but on elementary reasoning: if void were impossible, nature would have nothing to fear. The resistance that the slabs offer to being disconnected indicates that their separation leads to the formation of a void, even if only for a brief time. Like his Aristotelian opponents, Galileo calls "fear of the void" what was later to be called "air pressure." But this fear is in his view a positive physical force, the value of which can be measured by empirical means.

As we have seen, the existence of a void space, which Aristotle had rejected as a physical impossibility, was regarded by medieval authors as a possibility *de potentia Dei absoluta*: If God created a vacuum—Albert of Saxony argued—a body could move in it. But in the ordinary course of events this possibility was not realised, because nature would prevent the formation of a void. For Galileo, by contrast, it was essential to prove that

---

73 Galilei, *Two New Sciences*, 19–20.

74 Ibid., 20–21.

the existence of a vacuum represented a real physical possibility, albeit one that was not easily realised, since it was the vacuum that provided the theatre in which his new science of motion could be put on stage.

Conclusion

Galileo's theory of motion, which had been presented piecemeal in the *Dialogue*, was given an axiomatic form in the *Two New Sciences*. In the Third Day of this work, we encounter Salviati, who reads out aloud a Latin treatise entitled *On Naturally Accelerated Motion*, written by a friend Academician (Galileo) according to the model of Euclid's elements. As Bertoloni Meli has noticed, "the deductive structure of Galileo's science poses constraints on the role and presentation of experiments."[75] In the Third Day, one encounters in fact only a few experiments, which moreover do not play a foundational role, but rather serve to confirm established results.

But the treatise *On Naturally Accelerated Motion* only represents the *pars construens* of Galileo's science of motion. The *pars destruens*, which consists in a critique of the Aristotelian world-view, and more specifically in the demolition of the dichotomies super-lunar/sub-lunar, heavy/light, violent/natural, is to be found in the first two Days of the *Dialogue* and in the First Day of the *Two New Sciences*. These two texts, which are both written in an informal dialogical style, not only abound in *sensate esperienze*, but also repeatedly appeal to imaginary scenarios, which, as we have seen, are mostly derived from medieval sources.

I have tried to demonstrate that scholastic thought experiments serve Galileo to show that the conclusions of Aristotelian natural philosophers are based less on the study of natural occurrences than on the contemplation of imaginary scenarios. Moreover, the way in which Aristotelian philosophers analyse these scenarios is viciously circular:

> SALV.: Some men reasoning preposterously, first establish some conclusion in their mind which, either because of being their own or because of their having received it from some person who has their entire confidence, impresses them so deeply that one finds it impossible ever to get it out of their heads …
>
> SAGR.: I know; such men do not deduce their conclusion from its premises or establish it by reason, but they accommodate (I should have said dis-

---

[75] Bertoloni Meli, "The Role of Numerical Tables," 180.

commode and distort) the premises and reasons to a conclusion which for them is already established and nailed down.[76]

It must also be clear that the arguments that in this article have been lumped together under the label "thought experiments" were not regarded by Galileo as a unitary category. The imaginary situations invoked by the partisans of the *impetus* theory were, in his eyes, nothing else than a surrogate for real experience. They were paradigmatic cases, which could be analysed by analogy with everyday observations and which served to disclose some contradictions in the Aristotelian account of natural and violent motion. Galileo's purpose was to show that these cases, if re-examined in the light of the principles of his new science of motion, could be used to refute Aristotle, not only to correct him.

Galileo's attitude towards the scholastic arguments *de potentia Dei absoluta* was, as we have seen, far more critical. As Salviati reminded Simplicio in the *Dialogue*, natural philosophers could only say something meaningful about the ordinary course of nature. This is why he dismissed as ridiculous the habit of making predictions about situations brought about by God's miraculous intervention, as well as the pretence of distinguishing between what was the case in our world and what could have been the case in alternative worlds. By reinterpreting medieval cosmological thought experiments in the light of his new science of motion, Galileo showed that the laws ruling upon terrestrial bodies could be used to explain the behaviour of celestial ones. In his hands, the scholastic *secundum imaginationem* arguments became an instrument to simplify the *world*, rather than to multiply the *worlds*.

[76] Galilei, *Dialogue*, 321–322.

ON KANT'S CRITIQUE OF THOUGHT EXPERIMENTS IN EARLY MODERN PHILOSOPHY

Stelios Virvidakis

1. Kant's writings abound in thought experiments of various kinds, providing premises of arguments, or developing arguments that constitute parts of a lengthy line of reasoning. The most famous ones include the description of "incongruous counterparts" in his essay "Concerning the Ultimate Ground of the Differentiation of Directions in Space," in his Inaugural Dissertation and in the *Prolegomena to any Future Metaphysics*,[1] as well as those involving the imaginary conditions and states of affairs invoked in the metaphysical expositions of space and time in the Transcendental Aesthetic, in the Transcendental Deduction, in the Analogies of Experience and in the Antinomies in the Transcendental Dialectic of the *Critique of Pure Reason*,[2] and some of the hypothetical examples discussed in the *Critique of Practical Reason*.[3] In fact, Kant was preoccupied with the idea of mental experimentation in philosophy.

[1] See Kant, "Von dem ersten Grunde," *passim*; *De mundi sensibilis*, 403; Kant, *Prolegomena*, § 13.

[2] See the arguments in the metaphysical expositions of the concepts of space and time, and the passages about the possibility of irregular changes of the colour of cinnabar and, in natural phenomena, the weight of smoke arrived at after the subtraction of the weight of ashes from the weight of the wood burnt, the relation between the succession of perceptions and the succession of the positions of a ship moving downstream, and the arguments for the theses and antitheses in the first two Antinomies (Kant, *Critique of Pure Reason*: A22/B37-A25/B40, A30/B46-A32/B48, A100-A101, A185/ B228, A192/B237-A193/B238, A426/B454-A437/B465).

[3] See, for example, the hypothetical case of the man who is asked to give false testimony against an honourable man, on pain of immediate execution, and Kant's reference to thought experiments which allow us to grasp the clear distinction of the (empirical) principle of happiness, from the (rational) principle of morality, compared to the experiment of the (chemical) analyst, who "adds alkali to a solution of calcareous earth in hydrochloric acid, the acid at once releases the lime and unites with the alkali, and the lime is precipitated" (*Critique of Practical Reason*, 163–164, 214). One is struck by Kant's suggestion that here the philosopher cannot appeal to "any intuition (a pure *noumenon*) at the basis" [of his practical cognition], and yet, "*almost like a chemist*, can at any time set up an experiment with every human practical reason in order to distinguish the moral (pure) determining ground from the empirical ..." (ibid., 214, my emphasis).

Having proposed the central idea of his "Copernican turn" as a revolutionary assumption, already successfully tried in different ways in mathematics and in natural science, which might lead to progress also in metaphysics, he explicitly presented his entire critical enterprise as an extensive thought experiment aiming at a form of *reductio*, in a footnote of the preface to the second edition of the first *Critique*:

> This method, imitated from the method of those who study nature, thus consists in this: to seek the elements of pure reason in that which admits of being confirmed or refuted through an experiment. Now the propositions of pure reason, especially when they venture beyond all boundaries of possible experience, admit of no test by experiment with their objects (as in natural science): thus to experiment will be feasible only with the concepts and principles that we assume *a priori*, by arranging the latter so that the same objects can be considered from two different sides, on the one side as objects of the senses and the understanding for experience, and on the other side as objects that are merely thought at most for isolated reason striving beyond the bounds of experience. If we now find that there is agreement with the principle of pure reason when things are considered from this twofold standpoint, but that an unavoidable conflict of reason with itself arises with a single standpoint, then the experiment decides for the correctness of that distinction.[4]

Now, a proper account of Kant's thought experiments in his major works and an examination of their role in the development of his epistemology, of his critical metaphysics of experience and of his moral theory, would require a detailed study of complex issues pertaining to his transcendental strategy. The evaluation of their force would involve an assessment of the degree of their dependence on questionable idealist assumptions, presumably giving rise to legitimate doubts about what they really manage to establish. However, it is perhaps equally important for a good understanding of Kant's relevant methodological guidelines to focus on his strictures against certain well-known thought experiments of earlier philosophers. Thus, one may eventually be able to draw upon basic positions that are worth exploring, in so far as they cast light upon his own theory and practice of testing philosophical claims by appealing to imaginary situations. Moreover, we should not forget that Kant often deployed "counter-thought experiments" aimed at rebutting the thought experiments of his opponents.[5]

---

[4] Kant, *Critique of Pure Reason*, BXIXn.

[5] The term "counter-thought experiments" is used by J.R. Brown in his article "Thought Experiments" in the *Stanford Encyclopedia of Philosophy*. See Norton, "Why

Indeed, I do believe that Kant's negative approach towards some of the most interesting speculative hypotheses of his predecessors, supposedly leading to substantive philosophical conclusions, provides a basis for fruitful methodological reflections of a more general character. One may begin by turning to objections directed against particular thought experiments of pre-critical thinkers, inspired by the limitations and the constraints Kant was determined to impose on philosophical theorising as a whole. At a second stage, one should ponder precisely these limitations and constraints, following the main guidelines of his transcendental enterprise and conforming to his conception of possible experience. Finally, one could try to extrapolate from the examples taken into consideration and from their particular treatment, eventually drawing a lesson of Kantian inspiration, which could be extended to all thought experiments in philosophy, regardless of the area and of the topics dealt with. The results obtained in this way might prove quite helpful for an elaboration of our present methodological concerns, especially regarding the appraisal of thought experiments that sometimes appear far-fetched, implausible, and practically useless to us.

Hence, the analysis proposed in what follows does not amount to a simple reconstruction of familiar Kantian doctrines and arguments, elaborated in a new guise. Our aim is not so much exegetical or interpretative as philosophical. In so far as these Kantian doctrines and arguments are related to common worries about thought experiments and to ways of responding to such worries, it is definitely worth assessing their effectiveness in cases to which they were originally applied. However, our discussion aims at determining the extent to which a proper construal of the transcendental perspective adopted by Kant could indicate general, reliable criteria for the evaluation of thought experiments that would not require any appeal to dubious metaphysical commitments.

2. Before proceeding to a summary presentation of case studies concentrating on Kant's texts, we need to be as clear as possible about the concepts and issues in question. We should bear in mind that the very notion of a thought experiment is perhaps irredeemably vague. Nevertheless, for the purposes of our discussion, I propose that we content ourselves with the broad description of "a technique for testing a hypothesis by imagining a situation and what would be said about it (or, more rarely,

Thought Experiments," 45–49; Atkinson and Peijnenburg, "Thought Experiments Poor?," 308–315.

what would happen in it) ... often used by philosophers to argue for (or against) a hypothesis about the meaning or applicability of a concept."[6] In fact, if we do not object to the wide scope of this attempt at a definition, we may regard many instances of hypothetical or counterfactual reasoning as thought experiments. To be sure, it is not always easy to see the argument or the pattern of reasoning behind a hypothetical scenario, and a good deal of reconstruction may be required. In any case, going back to my introductory remarks about Kant's varied use of thought experiments in philosophy, I submit that we could speak of thought experiments which *introduce arguments*, by providing the grounds for sustaining one or more of their initial premises, as well as of longer arguments *involving or including one or more thought experiments* at some point.[7] The imagination must be engaged somehow, but not necessarily through original and fanciful lengthy fictional scenarios.

Here, it is important to remember some of the most common worries or objections I have alluded to in the previous paragraphs. Many thought experiments in philosophy are regarded, by those who reject them or at least express doubts about their epistemic reliability, as circular, unrealistic and inconclusive. They cannot be trusted as "intuition pumps," to use the metaphor introduced by Daniel Dennett. More particularly, one could dwell on a number of serious qualms, most of which are interrelated:

(a) The hypotheses put forth are selected on the basis of tacit epistemological or metaphysical assumptions, which are quite controversial and cannot be set aside when we interpret the thought experiments in question. Hence, we cannot trust the conclusions arrived at.
(b) Even if they do not rely on particularly controversial assumptions, thought experiments often turn out to be circular and, apparently, this is a vicious circularity that undermines the claims they are supposed to sustain.
(c) There is no agreement on the beliefs or principles presupposed, even when they don't seem to involve specific philosophical theories.[8]

---

[6] See Audi, *Cambridge Dictionary*, 918.

[7] As an example of the first kind one could cite Descartes' sceptical scenarios in the *Meditations*, while some of Kant's thought experiments mentioned above, such as the ones invoked in the Analogies of Experience or the Antinomies in the *Critique of Pure Reason*, constitute instances of the second kind. In fact, according to Goffi and Roux's narrower construal, we should count as genuine thought experiments only the ones that can be regarded as arguments, even if they are not just arguments.

[8] Here, I have in mind the problems of some of the "unsuccessful" thought experi-

(d) Some philosophers express doubts about the *status* and the authority of the modal intuitions supposedly revealed and defended through thought experiments and, especially, through the employment of thought experiments in particular areas, for example in moral philosophy.[9]
(e) Critics often argue that the imaginary scenarios constructed as thought experiments in order to help us explore metaphysical possibilities do not cast light on what is really possible for us.[10] To put it differently, such scenarios may disclose very remote possible worlds that are too distant to be relevant to us.

Presumably, successful philosophical thought experiments, if there are any, do not display such defects.

Now, we have already hinted at the suspicion that some of the above objections could be raised against Kant's own thought probes. For instance, even if the starting points of his imaginative mental constructions are, indeed, more or less philosophically neutral, it is doubtful whether their implications do show what was originally intended, unless one has smuggled in question-begging premises in the course of their elaboration. However, although we may have to come back to this issue at the end of our discussion, what we are interested in primarily is Kant's attitude towards thought experiments of pre-critical philosophers. In fact, the author of the *Critique of Pure Reason* is one of the first thinkers who have tried to argue, in a systematic and principled way, that mere conceivability or imaginability does not entail real possibility, thus sustaining a version of objection (e), and perhaps (d), above. The epistemological constraints he invoked as indispensable for the development of metaphysics of experience also provided him with the main criteria for an assessment of thought experiments proposed by other philosophers. I intend to focus on Kant's explicit or implicit comments on some of these well-known thought experiments elaborated by rationalist thinkers, mostly Leibniz and, to some extent, also Descartes, including an attempt at the defence of the principle of the identity of indiscernibles and the sceptical hypotheses of the *Meditations*. At this point, I shall refer only very briefly to the more or less indirect critique of empiricist examples and thought

ments discussed by Goffi and Roux in their contribution to this volume. However, as I have already suggested, I doubt whether they would agree with my conception of thought experiments, which they probably regard as too wide.

9 See, for example, Jamieson, "Method and Moral Theory," 476–487, 484–486.

10 See Wilkes, *Real People*.

experiments that Kant must have been familiar with, such as those employed by Hume, and I will not attempt to reflect on what Kant *might have said* about some of those he didn't address at all, such as Locke 's hypothetical scenarios bearing on the problem of personal identity. My remarks constitute only a preliminary part of work in progress focusing on a Kantian conception of thought experiments.

In any case, I am interested in assessing the plausibility of Kant's criteria underlying his qualms about pre-critical thought experiments, to the extent that they do not rely on an acceptance of the full implications of the Copernican turn and on the adoption of the central tenets of transcendental idealism. To be sure, we may still construe the methodological guidelines imposing such criteria as the expression of an indefensible idealist or verificationist dogma. However, if we are sympathetic to Kant's views, or rather to a variant or descendant of his views about the modalities involved in philosophical theorising, we may be willing to endorse them at least in part. Thus, we could perhaps be willing to agree on an extrapolation or hypothetical extension of the directives that we will have isolated to other thought experiments, elaborated in different contexts, regardless of the plausibility of one or another substantive form of transcendental idealism.

3. Indeed, Leibniz is the main target of Kant's analysis in the section "On the amphiboly of the concepts of reflection through a confusion of the empirical use of the understanding with the transcendental," presented as an appendix to the third chapter of the Analytic of Principles of the first *Critique*, dealing with "the ground of the distinction of all objects between Phenomena and Noumena."[11] The sage of Königsberg wanted to prove the inapplicability of Leibniz' principle of the identity of indiscernibles in the world of things regarded as phenomena, that is, as they appear to us and not as they are in themselves.[12] Leibniz, in his correspondence with Clarke, appealed to the principle, starting with a story that seems to involve real life empirical observation and asking us to concede that there cannot be any numerically distinct indiscernible objects, either in the realm of sensible or in that of insensible things. His apparently "empirical" story is extended into an imaginary supposition of "two

[11] Kant, *Critique of Pure Reason*: A260/B316—A292/B349.

[12] Here, I am glossing over the distinction between *appearances* (*Erscheinungen*) and *phenomena* ("appearances in so far as they are thought as objects according to the unity of the categories," A248), which is not important for the purposes of our discussion.

indiscernibles," "such as two pieces of matter perfectly alike," and to a thought experiment about the indiscernibility of insensible elements:

> 4. There is no such thing as two individuals indiscernible from each other. An ingenious gentleman of my acquaintance, discoursing with me, in the presence of her Electoral Highness the Princess Sophia, in the garden of Herrenhausen, thought he could find two leaves perfectly alike. The Princess defied him to do it, and he ran all over the garden a long time to look for some; but it was to no purpose. Two drops of water, or milk, viewed with a microscope, will appear distinguishable from each other. This is an argument against atoms; which are confuted, as well as a vacuum, by the principles of true metaphysics ... To suppose two things indiscernible is to suppose the same thing under two names. And to suppose that the universe could have had at first another position of time and place, than that which it actually had; and yet that all the parts of the universe should have had the same situation among themselves as that which they actually had; such a supposition, I say, is an impossible fiction.
>
> 21. The supposition of two indiscernibles, such as two pieces of matter perfectly alike, seems indeed to be possible in abstract terms; but it is not consistent with the order of things, nor with the divine wisdom, by which nothing is admitted without reason ...
>
> 23. I said, that in sensible things, two that are indiscernible from each other, can never be found; that (for instance) two leaves in a garden, or two drops of water, perfectly alike, are not to be found ...
>
> 24. I believe that these general observations in things sensible, hold also in proportion in things insensible ... And it is a great objection against indiscernibles that no instance of them is to be found. There is nothing simple, in my opinion, but true monads, which have neither parts nor extension. Simple bodies, and even perfectly similar ones, are a consequence of the false hypothesis of a vacuum and of atoms, or of lazy philosophy, which does not carry on the analysis of things, and fancies it can attain to the first material elements of nature, because our imagination would be therewith satisfied.
>
> 25. When I say that there are two drops of water perfectly alike, or any two other bodies indiscernible from each other; I don't say 'tis absolutely impossible to suppose them; but that 'tis a thing contrary to the divine wisdom, and which consequently does not exist.[13]

It must be admitted that the logical and theological reasons provided by Leibniz, not only in order to explain the fact that we do not actually find exactly indistinguishable things, but also with a view to justifying

[13] Mr. Leibniz's fourth paper—reply to Dr. Clarke, § 1–4,6 and Mr. Leibniz's fifth paper—reply to Dr. Clarke, § 21–25, in Alexander, *Correspondence*, 36–37, 39–40, 61–62.

the speculative hypothesis that however much we searched or considered insensible things, we would not be able to find indiscernibles, are not very convincing. Thus, his argument involves an appeal to the principle of sufficient reason and to the wisdom of God who created the world. It should be noted that Leibniz allowed for the "possibility in abstract terms" of the supposition of two distinct indiscernibles, but this *prima facie* logical possibility is immediately ruled out since its acceptance would imply "admitting something without reason" (§ 21, 25). The impossibility of the further imaginary supposition, "that the world could have had at first another position of time and place, than that which it actually had; and that all the parts of the universe should have had the same situation among themselves as that which they actually had," is also thought to be demonstrated through the emphasis on a necessary agreement with "the order of things" and with "divine reason."[14] Thus, for Leibniz also, conceivability or imaginability—'abstract possibility'—doesn't amount to real possibility, which, however, is understood in logical and metaphysical terms, not taking into account the necessary conditions of the possibility of experience as setting negative limits, but also as revealing positive grounds for the applicability of *a priori* concepts to the world as it appears to us.

If we wanted to engage in a detailed examination of the issues raised by these passages, we should eventually turn to Leibniz' conception of space and its relation to the picture of the world created by God in conformity to the principle of sufficient reason. We would then have to undertake an evaluation of Kant's response in relation to his account of space, articulated in the Transcendental Aesthetic, where he argued both against Leibniz and his Newtonian opponents, and proposed a third alternative. What we are more interested in here, however, is the fact that the Copernican viewpoint adopted by Kant allowed him to use position in space as an irreducible criterion of individuation that renders the principle of the identity of indiscernibles inapplicable as far as phenomena are concerned:

> E. Identity and Difference. If an object is presented to us several times, but always with the same inner determinations (*qualitas* et *quantitas*), then it is always exactly the same if it counts as an object of pure understanding, not many but only one thing (*numerica identitas*); but if it is appearance, then the issue is not the comparison of concepts, but rather however identical everything may be in regard to that, the difference of the places of these

[14] Ibid., 37.

appearances at the same time is still an adequate ground for the numerical difference of the object (of the senses) itself. Thus, in the case of two drops of water one can completely abstract from all inner difference (of quality and quantity) and it is enough that they be intuited in different places at the same time in order for them to be held to be numerically different. Leibniz took the appearances for things in themselves, thus for *intelligibilia*, i.e. objects of the pure understanding (although on account of the confusion of their representations he labelled them with the name of phenomena), and there his principle of non-discernibility (*principium identitatis indiscernibilium*) could surely not be disputed, but since they are objects of sensibility, and the understanding with regard to them is not of pure, but of empirical use, multiplicity and numerical difference are already given by space itself as the condition of outer appearances. For a part of space, even though it might be completely similar and equal to another, is nevertheless outside of it, and is on that account a different part from that which is added to it in order to constitute a larger space, and this must therefore hold of everything that exists simultaneously in the various positions in space, no matter how similar and equal they might otherwise be …'

F. The principle of indiscernibles is really based on the presupposition that if a certain distinction is not to be found in the concept of a thing in general, then it is also not to be found in the things themselves; consequently all things are completely identical, even numerically identical, that are not already distinguished from each other in their concepts (as to quality or quantity). But since in the mere concept of anything abstraction is made from many necessary conditions of an intuition, it is with particular haste that that from which abstraction has been made is taken as something that is not to be encountered at all, and nothing conceded to the thing except what is contained in the concept … The concept of a cubic foot of space, wherever and however often I think it, is in itself always completely the same. Yet two cubic feet are nevertheless distinguished in space merely through their locations numerically diverse; these are conditions of the intuition in which the object of this concept is given, which do not belong to the concept but to the entire sensibility … [15]

In other words, Kant is telling us that when it comes to thought experiments and suppositions about the phenomenal world, we cannot ignore the epistemic conditions of sensibility which involve space as a form of intuition and not just as a set of relations, an order of coexisting things. Here, we cannot limit or exclude real possibility merely on the basis of abstract logical considerations, apart from the fact that the logical and theological reasons apparently invoked by Leibniz are not very good ones. The more general claim elaborated in the Amphiboly section is that

[15] Kant, *Critique of Pure Reason*, A263/B319—A264/B320, A281/A282—B237/B238.

all such issues cannot be resolved unless one engages in transcendental reflection, and thus develops a "transcendental topic" or "topology," making possible the assignment to each representation of its proper transcendental place. The comparison of representations on the basis of the relations of "identity and difference," "agreement and opposition," "inner and outer" and "determinable and determination" (matter and form) must proceed *after* we realise whether we are dealing with representations subjected to the conditions of sensibility, that is the forms of intuition, or with just pure concepts of the understanding in their employment by reason attempting to talk about *noumena*.[16]

4. Similarly, when we turn to the Refutation of Idealism and to the Paralogisms of Transcendental Psychology we soon realise that both the residual worries of Descartes' hyperbolical doubts, the dreaming hypothesis and the evil genius hypothesis, which according to my broad conception do count as thought experiments, the contemporary descendants of which are Brain-in-vat scenarios, and the supposed substantive metaphysical results of the *Meditations* are unwarranted. Descartes, who, like Leibniz, ignored transcendental reflection and the epistemic conditions of his own representations, was unable to understand that the immediacy of his inner sense didn't provide him with privileged access to a unitary, continuous and immaterial self. Nor was he aware of the fact that the scenarios he was imagining, presumably allowing him to grasp with certainty his own existence as of a unitary, simple and continuous spiritual being, while casting doubt on the existence of the external world, underestimated the need for a permanent and regular dimension of outer substances as a necessary condition for the determination of the identity of his own self in time.

Thus, when Descartes concluded, in the second *Meditation*, following the *cogito* argument, that he is "a thinking thing," and a little after the thought experiment of the piece of wax, affirmed that there is "nothing that is easier for him to know than his mind,"[17] he was, according to Kant, guilty of the paralogisms of simplicity and personality and did not understand that he cannot attain any knowledge about the metaphysical

---

[16] Ibid., A263/B319—A268/B324.

[17] Descartes, *Meditations*, 139–199, 152–153, 157. In fact, it could perhaps be argued that the *cogito* inference itself, which cannot be reduced to a syllogism, relies on, or involves, an elementary thought experiment that pursues the reasoning behind the sceptical scenarios.

nature of the self behind the *cogito*. In fact, one of Kant's "counter-thought experiments" seems quite persuasive and doesn't derive its force from transcendental idealist premises:

> An elastic ball that strikes another one in a straight line communicates to the latter its whole motion, hence its whole state (if one looks at their position in space). Now assuming substances, on the analogy with such bodies, in which representations together with consciousness of them, flow from one to another, a whole series of these substances may be thought, of which the first would communicate its own state, together with its consciousness, to the second, which would communicate its own state, together with that of the previous substance, to a third substance, and this in turn would share the states of all previous ones together with their consciousness and its own. The last substance would thus be conscious of all the states of the previously altered substances as its own states, because these states would have been carried over to it, together with the consciousness of them; and in spite of this it would not have been the very same person in all these states.[18]

On the other hand, the sceptical doubts themselves concerning the existence of the external world, which emerge from the imaginary circumstances of the first *Meditation* and persist after the failure of the attempted proofs of the existence of God, giving rise to a "paralogism of ideality," according to the first edition of the *Critique*, could be regarded as totally unfounded. One would be ready to dismiss them, following the thrust of the central argument of the Refutation of Idealism in the second edition, to the effect that, "the mere, but empirically determined consciousness of my own existence proves the existence of objects in space outside me."[19]

Here, we will not try to assess the validity of the transcendental argument put forth by Kant in order to establish the need for a dimension of external objects as a necessary *a priori* condition of intelligible, internal self-awareness. What we should remark is just the eventual employment of transcendental argumentation,[20] in order to show the unreliability of thought experiments with sceptical intent, which do not take into account the *a priori* conditions of experience *intelligible to us*. A similar

---

[18] Kant, *Critique of Pure Reason*, A364.

[19] Ibid., B275.

[20] Here, I am not going to discuss the issue of the similarities and the differences between contemporary post-Strawsonian transcendental arguments and Kant's original transcendental deductions. In the following section, we will turn briefly to characteristics of transcendental proofs, according to Kant's methodological account, which may be relevant to the discussion of constraints on all successful thought experiments. Concerning these issues, see Stern, *Transcendental Arguments* and *Transcendental Arguments and Scepticism*; Glock, *Strawson and Kant*.

strategy allowed Kant to deal with Humean sceptical worries, whether they are based on the deliverances of introspection leading to a denial of the existence of a unitary self *of any kind* in the "theatre of the mind,"[21] or on hypotheses about the possible movements of billiard balls, undermining our natural beliefs about the causal connections of events in the natural world.[22]

5. What is common in Kant's treatment of all thought experiments of previous philosophers, adopting either a rationalist or an empiricist approach, is the concern for a proper account of the possibility of experience. The problem with the hypothetical and counterfactual scenarios proposed by Descartes, Leibniz, and Hume is not so much the fact that they tacitly adopt controversial logical and metaphysical principles, or that they presuppose what they want to establish. It is their inability to acknowledge the existence of "epistemic conditions"[23] of the operation of our mental faculties, which impose important limitations on our access to reality considered in itself, and on the application of our categories to the pursuit of philosophical knowledge. What they lack is an appreciation of the need for a correct interpretation and application of the categories of modality in any argument or thought experiment, which one could legitimately rely on in any inquiry into the metaphysics of experience. Indeed, the idea of the possibility of experience and of its conditions is the key to any philosophical enterprise following the guidelines of the *Critique*. Kant's investigation of this notion provides the basis for his novel, subtle and nuanced approach to modalities.

In any case, in order to understand Kant's conception of real possibility involved in his assessment of thought experiments one has to turn to his transcendental theory of modalities presented in the Postulates of Empirical Thinking in General:

1. Whatever agrees with the formal conditions of experience (in accordance with intuition and concepts) is possible.

---

[21] See Hume, *Human Nature*, 252–253. Kant's thought experiment about the transmission of motion in elastic balls seems to confirm the Humean observations, but only regarding the metaphysical substantial self and not Kant's conception of a transcendental self guaranteeing the unity of apperception.

[22] See Hume, *Human Understanding*, 28–30.

[23] Concerning the notion of epistemic conditions and its significance for a "two aspect," instead of a "two-world" interpretation of transcendental idealism, see Allison, *Kant's Transcendental Idealism*, 75–156.

2. That which is connected with the material conditions of experience (of sensation) is actual.
3. That whose connection with the actual is determined in accordance with general conditions of experience is (exists) necessarily.[24]

It has been objected that this distinction among the schematised modal categories cannot help us in our philosophical work, since it turns out that the three categories turn out to be coextensive: everything that is an object for us in the world of experience is in a sense possible, actual *and* necessary.[25] However, what could be retained is Kant's important insight that we cannot settle the issue of the possibility or impossibility of things, of whether an object corresponds to a concept merely through abstract analysis, which would prove that a concept is not self-contradictory, or that it does not violate a set of basic logical principles. Intuition is essential for a *schematised* possibility to be established. In fact, in the dense final paragraphs of the Amphiboly of concepts of reflection, Kant focuses on the "highest" concept of an object in general (taken problematically, leaving undecided whether it is something or nothing) and provides a table of four kinds of "impossible" objects:

1. To the concepts of all, many, and one there is opposed the concept of that which cancels everything out, i.e. none, and thus the object of a concept to which no intuition that can be given corresponds is = nothing, i.e. a concept without an object, like the *noumena*, which cannot be counted among the possibilities although they must not on that ground be asserted to be impossible (*ens rationis*), or like something such as certain new fundamental forces, which one thinks, without contradiction, to be sure, but also without any example from experience even being thought, and which must therefore not be counted among the possibilities. [Empty concept without object].
2. Reality is something, negation is nothing, namely, a concept of the absence of an object, such as a shadow or cold (*nihil privativum*). [Empty object of a concept].
3. The mere form of an intuition, without substance, is in itself not an object, but the mere formal condition of one (as appearance),

[24] Kant, *Critique of Pure Reason*, A217-A218/B265-B266.

[25] For a first discussion of the philosophical difficulties besetting the construal of this account of the modal categories in the Postulates of Empirical Thought in General, see Milmed, "Possible Experience," 301–321 and Gardner, *Kant*, 196–198.

like pure space and pure time, which are to be sure something, as the forms for intuiting, but are not in themselves objects that are intuited (*ens imaginarium*) [Empty intuition without an object].

4. The object of a concept that contradicts itself is nothing because the concept is nothing, the impossible, like a rectilinear figure with two sides (*nihil negativum*) [Empty object without concept].

> ... One sees that the thought-entity (No. 1) is distinguished from the non-entity (No. 4) by the fact that the former may not be counted among the possibilities because it is a mere invention (although not self-contradictory), whereas the latter is opposed to possibility because even its concept cancels itself out. Both, however, are empty concepts. The *nihil privativum* (No. 2) and the *ens imaginarium* (No. 3), on the contrary, are empty *data* for concepts. If light were not given to the senses, then one would also not be able to represent darkness, and if extended beings were not perceived, one would not be able to represent space. Negation, as well as the mere form of intuition are, without something real, not objects.[26]

Hence, whenever we want to rule out certain possibilities we must be certain about what we are talking about and about the sense in which we decide that something is or is not possible. More generally, once we are able to distinguish between mere *logical* possibility, which is relevant to consistent *thought* about the *entia rationis* of the world considered in itself, but doesn't amount to *knowledge* about this *noumenal* aspect, and *real* or *transcendental* possibility, which is essential for reaching conclusions about the world as it appears to us, we know which thought experiments we should trust, in principle, and which we should not. Presumably, *transcendental* possibility would be more narrow than *logical*, but wider than mere *physical* possibility. Philosophers, who do not respect such a distinction of kinds, or ranges of modality, will construct all kinds of counterfactual hypotheses, trying in vain to establish metaphysical truths about what the world in itself must or could be like. Their thought experiments will continuously clash with the thought experiments of their opponents who reach contrary conclusions, and it would prove impossible for their debates to be adjudicated. Kant attempted to make this clear and to defend his critical diagnosis in the "Antinomies" section of the "Transcendental Dialectic." Thus,

---

[26] Kant, *Critique of Pure Reason*, A290/B347-A292/B349.

> A transcendental hypothesis, in which a mere idea of reason would be used for the explanation of things in nature, would be thus no explanation at all, since that which one does not adequately understand on the basis of known empirical principles, would be explained by means of something about which one understands nothing at all. And the principle of such a hypothesis would really serve only for the satisfaction of reason and not for the advancement of the use of the understanding in regard to objects.[27]

What pre-critical thinkers fail to understand is that their quest for modal intuitions supposedly leading to metaphysical cognition, or, in the case of sceptical empiricists, to a negative verdict about its attainability, should first be subjected to a transcendental scrutiny. Only after such a scrutiny should we attempt thought experiments, which would help us establish modest ontological conclusions, pertaining only to the *a priori* structure of the world as it appears to us and constituting the basis of metaphysics of experience. The thought experiments that interest us would be the ones helping us to explore possible worlds that are *sufficiently close* to our world to be intelligible to us and to beings like us. These may be thought experiments involved in or connected to transcendental argumentation.

Now, Kant, in the Transcendental Methodology of the *Critique of Pure Reason*, provided directives for the kind of philosophical work, which he considered to be legitimate and promising. One could undertake a careful study of this section with a view to elaborating a transcendental theory of the main rules for the construction of useful thought experiments and for the critique of misguided ones, from Leibniz' hypothetical generalisation of his example of the two leaves or drops or water, to Hume's billiard balls that were supposed to be able to move in any conceivable direction in the world of intelligible appearances. Here, we shall only mention one of the most fundamental rules, formulated in the section of the "Discipline of Pure Reason" with respect to its proofs:

> The first rule, therefore is this: to attempt no transcendental proofs without having first considered whence one can justifiably derive the principles on which one intends to build and with what right one can expect success in inferences from them.[28]

Actually, I would like to close my account of Kant's approach by suggesting that this rule, interpreted in a broad sense, not necessarily tied to the

[27] Ibid., A772/B800.

[28] Kant, *Critique of Pure Reason*, A786/B814. Here, one could perhaps replace "transcendental proofs" by "thought experiments." It should be noted, however, that Kant's methodological rules are rather obscure at many points and may give rise to serious queries and objections.

Copernican framework, could perhaps prove useful as a methodological maxim for all attempts at a critical evaluation of thought experiments. We could resort to it whenever we feel we can't rely on thought experiments we are asked to engage in, because we suspect one or more of the factors identified in the objections that we summarised at the beginning of our analysis may be present.

Before concluding, however, we have to acknowledge the importance of issues which we cannot discuss in this paper. What if, as we have already intimated, we do believe that the critical restrictions imposed on philosophical investigations by the transcendental enterprise amount to an excessive idealist or verificationist limitation of the scope of philosophising?[29] How should we apply the constraints entailed by the notion of the epistemic conditions of possible experience in the area of moral philosophy? Notoriously, Kant thought that his approach in the realm of practical reason does not have to conform to the limits of the critical metaphysics of the world as it appears to us, and when we deal with worries about thought experiments and the modalities they rely on in this domain.[30] We may perhaps have to look for different, *practical*, rather than *epistemic* constraints.

I confess I don't have a good answer to such questions. I share the worries of many contemporary philosophers about the uncontrolled and uncritical use of many thought experiments and I do hope that the legacy of the Kantian transcendental approach may provide a useful corrective. The pursuit of further inquiry into the possible uses of the approach to thought experiments that I have tried to sketch would be the task of a long-term research project.[31]

[29] An example of such an unwarranted limitation is the diagnosis of a kind of indulgence in accusations of nonsensicality addressed to partisans of metaphysical theorising by Wittgensteinian philosophers, who consider thought experiments mostly as devices for exposing the limitations of our language-games or forms of life and for bringing to light the absurdity of metaphysical theorising.

[30] See Jamieson, "Method and Moral Theory."

[31] I would like to thank the participants in the workshops on thought experiments where earlier versions of this paper were presented for their questions and critical remarks, and especially Pascal Engel for his extensive comments and his anti-Kantian objections.

PART THREE

HOW DO THOUGHT EXPERIMENTS WORK?

PHILOSOPHICAL THOUGHT EXPERIMENTS: IN OR OUT OF THE ARMCHAIR?

Pascal Engel

1. *Introduction: Philosophical Thought Experiments*

It is not easy to say what a thought experiment in science is and what role it is supposed to play: is it really an experiment or is it merely a thought exercise? There are, however, clear examples: Galileo's reasoning about motion, Newton's bucket, Einstein's lift, Schrödinger's cat, etc. Thought experiments seem to have their proper home in philosophy, widely held to be a discipline dealing with concepts. But in philosophy, the examples are much less clear-cut. Paradigmatic thought experiments are Locke's on personal identity, Strawson's auditory world in the second chapter of Individuals, Putnam's Twin Earth, Jackson's Mary the colour scientist, or Searle's Chinese room.[1] However, there are more borderline cases: Theseus' ship, Molyneux's problem, Descartes' piece-of-wax example in the Second Meditation, the Gettier cases, and all the machinery of examples invented by contemporary analytic philosophers for the discussion of particular issues in ethics, metaphysics, philosophy of mind, or epistemology.[2] It is not clear whether these are examples of thought experiments because they involve our imagination or because they involve a certain kind of reasoning. For instance, is the hypothetical supposition of a state of nature in classical social contract theories a thought experiment? Are medieval reasonings about God and angels thought experiments? We tend to think that they are because we consider these creatures—or at least some of them—to be fictitious, but medieval philosophers did not think of them in this way. It is also an open question whether we should conceive of scientific thought experiments as experiments, which one could really have performed had the conditions been met, or whether they are meant to be forever imaginary. In

[1] There is a nice stage setting of these thought experiments in David Lodge's satire of artificial intelligence academia in *Thinks*, Penguin 2002.

[2] For a good survey and discussion, see Sorensen, *Thought Experiments*.

contrast, in the case of science, we have an idea of what an experiment is. The problem with philosophical thought experiments is that they do not seem to be, even remotely, experiments at all. They are, however, supposed to play an important role in philosophical argument and theories. What kind of role? For many contemporary philosophers, their role is to test, in thought, various intuitions that are supposed to confirm or to weaken certain philosophical claims. The epistemological problem is thus: what is the status of these intuitions and of the kind of information or knowledge—if such there be—they are supposed to provide? In this article, after having reviewed some main conceptions of scientific thought experiments, I shall mainly discuss the view, recently advocated by Timothy Williamson in *The Philosophy of Philosophy*, according to which a philosophical thought experiment is just a piece of ordinary counterfactual reasoning.

2. *Conceptions of Thought Experiments*

Ernst Mach famously defined *Gedankenexperimente* as the capacity to "imagine mentally the variation of facts." He said that this activity is not only available to the scientist, but also to the philosopher, the novelist, and the engineer. But what is their common form?

As the variety of examples given above suggests, it is not clear whether they have a common form. But it is at least plausible to suggest that thought experiments have the following structure: to devise a thought experiment involves the conception of a possible situation against which we test our intuitions, and from which we reason about an actual case which we contrast with the possible one. So there is, at the basis of a thought experiment, a modal claim to the effect that such and such a situation is possible. Possibility, however, can be understood in different ways. We can talk of physical possibility, of metaphysical possibility, of epistemic possibility, or of conceptual possibility. It would be circular or question-begging to say that physical thought experiments deal with physical possibilities, metaphysical thought experiments with metaphysical possibilities, conceptual thought experiments with concepts. For the question raised by most thought experiments is precisely the question of what kind of possibility claim is involved. In principle, a physical thought experiment is supposed to deal with our capacity to conceive of physical possibilities, and we evaluate these in the light of what we already know about the physical world. However, many physical thought experiments precisely involve going beyond what we already know about physical possibilities, or considering whether the envisaged situation does not contradict our existing knowledge of physical facts. They imply an extension of our conception or imagination to new possibilities, and the question is whether these possibilities are genuine ones. However, in order to answer the question whether a given situation is possible or not, we first have to answer another one: how can we have access to genuine possibilities? The latter question is epistemological; it asks what sort of faculty or cognitive capacity gives us access to the possible situation: Imagination? Conceptual understanding? Intuition? A priori reasoning? Empirical reasoning? These are not necessarily the same. So the modal question and the epistemological question are closely associated. For if we answer that the envisaged possibility is just a fiction of our imagination, we reject thought experiments as "mere" exercises of imagination, and not as a genuine possibility. This problem is also involved in the question raised by Mach, and answered positively by him, of whether thought experiments are a limiting case of real experiments: if they are real experiments, then the possibilities in question are indeed genuine possibilities. So the main issue about any thought experiment is: does conceivability imply possibility?[3] Much of the success of the method of thought experiments in any domain—physics, engineering, philosophy, or literature, to name the domains considered by Mach—depends upon how one answers this question.

Let us try to illustrate the proposal that a thought experiment involves a modal claim and an epistemological claim by giving some examples. Take first Lucretius' famous thought experiment in *De rerum natura*, which asks us to imagine an archer at the edge of the universe throwing a spear at it. If the spear bounces back, there is a wall beyond the limit, which is itself in space, hence in a limitless medium; and if it does not, there is no boundary; hence in both cases the conclusion is that the universe is infinite. This is a clear case in which we try to construct a possible situation in order to test a certain proposition, and in which the result is that the situation is not possible at all. The reasoning has the following structure:

1. It is possible that an archer throws a spear at the edge of the universe [modal claim];

[3] For a recent analysis of this classical problem (Descartes, Hume), see in particular Gendler and Hawthorne, *Conceivability and Possibility*.

2. if in this situation the spear bounces back, there is a wall outside the limit, and if it does not, there is no limit [consequence of (1)];
3. hence in either case there is no limit to the universe.

Lucretius' thought experiment is a reductio ad absurdum reasoning based on a hypothesis. The question left implicit is whether the hypothesis is a product of our imagination, or of our conceptual thought. In the case envisaged it involves imagination, but it is also clearly a thought experiment about our concept of space.

As a second example, take Frank Cameron Jackson's famous contemporary thought experiment about Mary the scientist, which is obviously reminiscent of Molyneux's question. Mary is a colour scientist who knows everything about the physics and the neurophysiology of colours, but who, having lived in a black and white room with a black and white television screen all her life, has never actually seen a coloured object. One day she is taken outside and given a coloured screen. Will she learn anything about colours? The structure of the thought experiment is the following:

1. It is possible that there is a person like Mary, who knows everything physical about colour but has never experienced colour [modal claim];
2. if Mary could experience colour, she would learn something new;
3. hence phenomenal properties are distinct from physical properties.

Here, as in the previous case, we are asked to react to a possible situation by using our intuition in order to validate a certain kind of reasoning. So the question arises about the nature and the source of this intuition.

If we consider the answers, which have been given to the latter question for thought experiments in science, there are four main options:

i. An empiricist account, according to which thought experiments are reasonings or arguments based on empirical premises, aimed at either deriving some contradiction within a theory, or at proving certain consequences of a hypothesis, or at making an inference to the best explanation. In this case, defended most explicitly by John Norton, there is no difference between a thought experiment and an ordinary reasoning from premises to conclusion, and one criticises a thought experiment just as one can criticise an argument. As Norton says: "Thought experiments are arguments which posit hypothetical or counterfactual states of affairs and which invoke particulars irrelevant to the generality of the conclusion."[4] The only thing the thought experiment adds is a contingent illustration, which could have been different. For instance, Schrödinger's cat could have been a dog, and the lift could have been a spaceship. So thought experiments are dispensable.

ii. A Platonistic account, according to which thought experiments are "telescopes pointed onto the world of abstract ideas," which give us access, through some sort of intuition of essences, to a world of Platonic entities.[5]

iii. An imaginarist account, according to which thought experiments do not really bring knowledge, but aim at producing, through the work of the imagination, heuristical fiction, which extends our representational capacities and measures the limits of our ordinary conceptual scheme.[6]

iv. A Wittgensteinian account, according to which thought experiments only test conceptual possibilities and are variations on the extension of our concepts.

However, it is clear that Norton's "deflationary" view, according to which there is nothing more to thoughts experiments in science than in ordinary empirical reasoning, is less easy to transpose to philosophy, since philosophy is usually thought of as a form of conceptual and non-empirical kind of thinking. But is it so clear that philosophy is wholly conceptual? If not, we can find parallel views about the epistemology of philosophical thought experiments and of the kind of modal knowledge that they involve.

3. *The Status of "Intuitions"*

The idea that we test philosophical claims and philosophical thought experiments on the basis of our "intuitions" is ubiquitous, but it is also ambiguous. The closest conception to James Robert Brown's view of thought experiments for philosophical intuitions consists in conceiving them as intellectual analogues of sense perception, which are the basis of a kind of *a priori* knowledge. The strongest *a priorist* view is defended

[4] Norton, "Why Thought Experiments."
[5] Brown, *Laboratory*.
[6] Gendler, "Galileo."

by George Bealer, who advocates what he calls "modest rationalism."[7] The view is considered rationalist because it says that our knowledge of necessities and possibilities is based on intuitions, which are independent of experience and distinct from ordinary beliefs. Intuitions, in Bealer's view, are intellectual "seemings" distinct from propositional attitudes such as beliefs: you can believe things that you do not intuit (e.g. that Rome is the capital of Italy), and you can intuit things that you do not believe (e.g. the axioms of naïve set theory). Bealer, however, does not take intellectual intuitions as infallible. On the contrary, they are fallible, and subject to be proven wrong. In other words, they are merely *prima facie* justified, although they have to be "modally reliable," that is stable, in order to count as evidence. This is why his rationalism counts as moderate.[8] So, according to such a conception, our intuitive answers to Lucretius' thought experiment count as stable, hence as *prima facie* correct answers to it.

At the opposite end of the spectrum of views on intuitions lies the ultra empiricist thesis defended by Stephen Stich (*Fragmentation*) and his associates (Nichols et al., "Meta-Skepticism"), for whom intuitions are just empirical judgements.[9] What Stich calls "epistemic romanticism" is the view that the intuitions elicited by philosophical thought experiments reflect universal and constant epistemic norms. According to Stich, however, what the "romantic" calls "intuitions" are actually mere empirical beliefs, which vary greatly from culture to culture, and from one socio-economic group to another. Hence so-called intuitions reflect no universal norm, but only variable representations. Stich uses classical philosophical cases, such as Gettier examples,[10] in order to show "experimentally" that the intuitions that they elicit are very different, depending upon whether they come from European/American or from Asian subjects. Gettier examples are standardly used to show that knowledge is not justified true belief. The experimental set-up consists of presenting Gettier cases such as the following to both Asian and European subjects:

> Bob has a friend, Jill, who has driven a Buick for many years. Bob therefore thinks that Jill drives an American car. He is not aware, however, that her

[7] Bealer, "Modal Epistemology."

[8] Bonjour, *In Defense*, and Peacocke, *The Realm*, hold related but distinct views on this point.

[9] Stich, *Fragmentation*; Nichols et al., "Meta-Skepticism."

[10] Gettier, "Is Justified."

> Buick has recently been stolen, and he is also not aware that Jill has replaced it with a Pontiac, which is a different kind of American car. Does Bob really know that Jill drives an American car, or does he only believe it?[11]

European/American subjects usually answer: Bob only *believes* that Jill has an American car; whereas Asian subjects say that Bob really *knows* the proposition. Stich gives all sorts of examples where the intuitions of Asians and Europeans about knowledge are subject to strong variations, and concludes that Gettier cases in no way show that knowledge is not justified true belief, since Asians accept that accidentally true beliefs might be knowledge.[12] His results, however, have been contested do they show that intuitions about *knowledge* vary from culture to culture, or do they show that the *social attitudes* and the concepts *associated* to knowledge vary from culture to culture? It is not clear that they show anything about the concept of *knowledge*.[13]

The whole debate—*a priori* intuitions or empirical beliefs—seems moot unless one does not clarify the kind of epistemic access one has to the situations described by thought experiments in philosophy. For this we have to understand the nature of the modal claim involved in philosophical thought experiments. What does our knowledge of possibility consist of?

4. *Williamson on Philosophical Thought Experiments and Counterfactuals*

From our prima facie characterisation of thought experiments in section 2, we can derive a simple answer: thought experiments are based on counterfactual reasoning. They consist of reasoning based on a "what if?" supposition or on counterfactual terms. The antecedents of counterfactuals describe a possible situation, the consequent describes a situation which we take to be compatible with the consequent when we evaluate counterfactuals as true: "If one threw a spear at the limit of the universe, such and such will happen"; "If Mary experienced colour, she would learn something." Lewis suggested that fictional discourse is a species of

[11] Nichols et al., "Meta-Skepticism."

[12] For similar results about many domains of Asian vs. Occidental thought, see Nisbett, *The Geography of Thought*, which I have criticised in Engel, "Is there a Geography of Thought."

[13] Sosa, "A Defence of the Use"; Engel, "Des avantages."

counterfactual reasoning.[14] Williamson has proposed the thesis that there is no more to our modal knowledge of possible facts than our capacity to handle and evaluate counterfactuals. He bases his account on the familiar equivalence between modalities and counterfactuals defined by Stalnaker in "A Theory of Conditionals":

(a) "A is necessary" = "If A were not the case, A would be the case"
$\Box A \equiv (\neg A \Box\!\!\rightarrow A)$
(b) "A is possible" = "It is not the case that if A were the case, A would not be the case"
$\Diamond A \equiv \neg(A \Box\!\!\rightarrow \neg A)$

From these equivalences, according to Williamson, we can argue that there is no more and no less to our imagining or conceiving possible cases in thought experiments than to our ordinary reasoning with counterfactuals. Counterfactual reasoning is quite a common form of reasoning not resting upon any faculty of intuition of metaphysical possibilities. Many counterfactual reasonings rely upon simple knowledge of empirical regularities. For instance, we say, "if the bush had not been there, the stone would have ended on the mountain path," or "if there had been an earthquake here, there would have been a tsunami," or "if I had been at the airport five minutes earlier, I would not have missed my flight." Others involving more work of imagination are more difficult to assess: "If Hitler had invaded England in 1940, he would have won the war"; "If I had been Audrey Hepburn, I would have been slim." Some do not involve imagination at all and are purely tautological: "If we had been one more person for dinner, we would have been 13." According to Williamson, just as we use counterfactuals to reason about possible states of affairs, we can use thought experiments to reason about possible state of affairs. But this kind of reasoning does not presuppose any use of a special faculty of intuition. Neither do we use a special sense of metaphysical possibility. We use usual cognitive resources: our empirical generalisations and inferences. So there is no more to the epistemology of thought experiments than to the epistemology of counterfactuals.

How can we apply this idea to the epistemology of thought experiments? It is not enough to remark that we can rephrase familiar thought experiments in counterfactual terms such as:

---

[14] Lewis, "Truth in Fiction."

1. If my left brain hemisphere had been transplanted into Righty's brain, he would have had the same emotions as me;
2. If Theseus' ship had been replaced plank by plank, it would have been the same ship;
3. If my world were purely auditory, I would have to rely on the pitch of sounds to navigate through space;
4. If Condillac's statue had only touch and smell, it could not perceive far distances etc.

In order to see what kind of work a counterfactual analysis does, we have to look more closely at the structure of one philosophical thought experiment. Williamson takes the example of the Gettier cases.

It is not immediately apparent that in contemporary epistemology Gettier cases are cases of philosophical experiments. As we have seen, they start from the notion that a necessary and sufficient condition for knowledge is justified true belief, and give counterexamples to this condition by exploiting the elementary logical point that some logical consequences of falsehood are truths, and the elementary epistemological point that deduction is a way of transmitting justification from the premises to the conclusion of an argument. A subject has a justified belief in a false premise q, deduces from it competently a true belief p, and believes p on that basis. So the reasoning starts from the Gettier condition: justified true belief is a necessary and sufficient condition on knowledge, which can translate thus:

(1) $\forall xp\ (K(x,p) \equiv JTB(x,p))$

The objection to (1) is the presentation of a Gettier case: it is possible that one has a justified true belief that p which is not knowledge.

(2) $\Diamond \exists x\ \exists p\ GC(x,p)$

Now, what plays the role in the reasoning of the modal claim that I had identified above is the following possibility: if the Gettier case had occurred, then the subject would have had a justified true belief in p without knowing, i.e.

(3) $\Diamond \exists x\ \exists p\ GC(x,p) \;\Box\!\!\rightarrow\; \forall x\ \forall p\ (GC(x,p) \supset (JTB(x,p)\ \&\ \neg\ K(x,p)))$ (where "$\Box\!\!\rightarrow$" denotes the counterfactual conditional "if it were the case that ...")

Hence JTB without K is possible:

(4) $\Diamond xp\ (JTB(x,p)\ \&\ \neg\ K(x,p))$

But (4) contradicts (1).[15]

The modality here is essential. For the Gettier case is imaginary and considered as possible. It is not a counterexample to the non-modal claim that in fact every case of knowledge is a case of justified true belief, and vice versa.[16] Gettier can claim that his case is possible, as in (2), not that it is actual. Since the "possibly" qualification is essential in (2), the "necessarily" qualification is essential in (1) if the objection is to stand.

So on this analysis, a thought experiment such as Gettier's is a modal argument going from a modal premise to a modal conclusion. It is the role of imagination to verify the premises. The major premise (3) is a counterfactual conditional. So there is nothing special about intuitions concerning cases in standard analytic epistemology. It does not rely on a special kind of faculty of insight, which would be a priori. It simply relies on ordinary reasoning about counterfactuals, which involves nothing but empirical knowledge.

5. *Replies to Some Objections to the Counterfactuality Thesis*

Williamson's counterfactuality thesis about thought experiments has strong similarities with the kind of deflationary view about thought experiments advocated by writers like Norton.[17] Williamson claims that we do not need any spatial faculty of insight into metaphysical possibilities to assess thought experiments, but only the humdrum cognitive resources involved in counterfactual reasoning. Nor do they rely on a specific conceptual faculty of intuition that would be specific to philosophy. In this respect, they do not rest upon any kind of a priori knowledge. Actually, the Gettier cases have nothing far-fetched, and, as Williamson notes, one can construct real life examples of Gettier cases. For instance, I can tell you that I have lectured in Algeria, and believing me, you can infer justifiably that I have lectured in North Africa. That belief is true, but the initial one was false: I have never lectured in Algeria, only in Tunisia.

There are some prima facie objections to Williamson's thesis. First of all, can the Stalnaker equivalences (a) and (b) above show that, when we understand statements about possibility, we do not understand anything more than the equivalent counterfactual statements? The fact that the

[15] Williamson, "Armchair Philosophy"; *Philosophy of Philosophy*, 183–187.
[16] Williamson, *Philosophy of Philosophy*, 185.
[17] Norton, "Why Thought Experiments."

Stalnaker thesis of the reduction of modalities to counterfactuals holds does not imply that the epistemology of modalities reduces to the epistemology of counterfactuals. To take another example of logical equivalence, the fact that (¬A & ¬B) is logically equivalent to ¬(A∨B) does not imply that the epistemology of conjunction is the same as the epistemology of disjunction; not any more than when the grocer tells me that he has neither jam nor quamkats, I need to understand that it is not the case that he either has jams or quamkats. We do not translate automatically the "necessary" in

> Socrates is necessarily human

into a counterfactual:

> If Socrates had not been human, 2 + 2 would have been 5.

But the objection misfires. The equivalences (a) and (b) do not tell us that we know modal facts by knowing counterfactual conditionals. At best they tell us that it is a way of knowing modal facts. So the logical equivalences (a) and (b) do not prove the epistemological equivalence of modal knowledge and counterfactual knowledge. It is only supposed to make plausible the claim that our intuitions of possibility and necessity are closely associated with our intuitions about counterfactuals.

Another objection is that Williamson's analysis cannot show that our modal knowledge of the premises of thought experiments (such as (3) of the Gettier reasoning above) is counterfactual knowledge, since the latter presupposes the former. We use our intuitions about necessity to handle counterfactual claims. Indeed, this is what David Lewis' (Counterfactuals) analysis of counterfactuals in terms of possible worlds and similarity spheres within possible worlds does.[18] When philosophical thought experiments involve far-fetched possibilities, for instance thought experiments about personal identity concerning science fiction cases of teleportation or half brain exchanges, what help can we get from the rephrasing of the thought experiments into counterfactual claims?

This objection, however, rests upon a misunderstanding: the counterfactuality claim is not that counterfactuals are prior to necessity and possibility claims, but that they come together. The counterfactuality thesis does not say that we understand certain philosophical thought experiments about remote possibilities more easily when we translate them

[18] Lewis, *Counterfactuals*.

into counterfactual terms. It says that we understand them neither better nor worse. In other words, if a thought experiment is unrealistic, it will remain so.

A third objection to the counterfactuality thesis is that it presupposes that counterfactual conditionals have truth conditions and can be evaluated truth conditionally. However, this thesis is rejected by those who adhere to the probabilistic analysis of counterfactuals, according to which these have only assertion conditions in terms of the conditional probability of the consequent given the antecedent.[19] But apart from the fact that this analysis is generally taken to work better for indicative than for subjunctive conditionals, it would not fit the structure of thought experiment reasoning, which involves the truth of propositions about possible cases, not their probability.

There is another, subtler, reason why the suggestion would be wrong. Typically, if one understands counterfactuals in terms of degrees of subjective probability, or in some analysis in terms of the familiar Ramseyan idea that the antecedent consists in adding a belief to our stock of beliefs and we see whether the consequent accords with it, the kind of possibility that we are dealing with is epistemic possibility (understood in degree or belief or in subjective probability terms). But this would result in making all thought experiments merely reasonings about epistemic possibilities. But some thought experiments—many actually—deal with metaphysical possibilities. Clearly, Galileo's thought experiments, or Lucretius,' do not deal with whether we can think certain motions as possible, nor with the subjective probability that we grant the hypothesis that the propositions describing them could be true. To limit the scope of thought experimental thinking to epistemic possibilities would have the effects of transforming our modal question into a question about our thoughts and beliefs. No doubt a fictionalist, or a conceptualist—Kantian about modalities would probably welcome this idea, which certainly fits some thought experiments (in particular Strawson's auditory world), but there is no reason to accept it in the first place. But it would beg our main question—what kind of modality is involved in thought experiments—to adopt this kind of answer. If we want to take into account the whole range of thought experiments, we had better not presuppose that they deal only with epistemic possibilities.

[19] Adams, *Logic of Conditionals*.

A fourth objection concerns the nature of cognitive resources involved in thought experiments. The counterfactuality thesis says nothing about them. It does not tell us whether our cognitive faculties in evaluating counterfactual claims rely on imagination or on conceptual resources, or simply on empirical beliefs. For instance, one might give an analysis of counterfactual reasoning in terms of a simulationist analysis of imagination[20] or in terms of mental models.[21] But the counterfactuality thesis does not pretend to decide which of these views is correct. Since it holds that the cognitive resources involved in thought experiments are just those of our ordinary thinking about counterfactuals, it presupposes that any correct story about how we understand counterfactuals will be useful for understanding thought experiments.

Finally, one might object that Williamson's analysis of the Gettier example is fairly restricted and might not extend to other philosophical thought experiments. But if one agrees with the analysis given above of the Lucretius and the Mary thought experiments, the modal claim is quite pervasive. In particular, it is quite important to see that they involve possibilities. Thought experiments do not propose empirical generalisations that could be tested and which could receive counterexamples. It is quite important here not to confound the usual practice of analytic philosophy of giving counterexamples with the practice of constructing thought experiments. Russell famously said about his analysis of definite descriptions that counterexamples in philosophy play the role that experiments play in science. Thus a counterexample to Russell's analysis of definite descriptions is a counterexample about a generalisation ("all phrases of the form 'the so and so' have the form 'there is a unique x which is so and so' "). But it is not a counterexample describing a possible case of a phrase of the form "the so and so" which is not analysable in Russellian terms. It is an actual case. In other words, thought experiments are not empirical generalisations, which could be disproved by actual cases. They are, if Williamson's analysis is right—and I think it is—about possibilities and necessities.

---

[20] Currie and Ravenscroft, *Recreative Minds.*

[21] Byrne, *Rational Imagination.*

6. *Metaphysical vs. Conceptual Possibilities*

A more convincing objection to Williamson's analysis is that it has the consequence that no genuine thought experiment bears upon our concepts or about our understanding of concepts. All thought experiments, in Williamson's view, bear on metaphysical possibilities: they are about the world, not about the way we think of it. But this claim seems very implausible, at least in regard to philosophical thought experiments. For many of them can indeed be understood as bearing upon our concepts, not upon the world. The common wisdom about, for instance, thought experiments about personal identity is that they test our common concept of personal identity. Strawson's purely auditory world in chapter II of Individuals bears clearly on our concept of an objective world: his question is whether such a concept would be coherent if our experience were limited to a world of sounds only. Many thought experiments explore the limits of the extension of our concepts, whether ordinary or philosophical, and are devised in order to test their coherence. Indeed, one dominant conception of philosophical analysis is that it deals with concepts, not things in themselves.[22] Thought experiments in this sense serve as a kind of critical tool. This does not preclude thought experiments from being about things and about possible states of affairs, but most of the time we have access to them only through our concepts. Thought experiments about persons or about freedom could hardly be about persons or freedom as themselves, they are first and foremost about our concepts of person and of freedom, even though they are supposed to show us something about persons and freedom. Many philosophical concepts involve theories, or at least a cluster of other concepts associated to them.

Williamson's analysis of the Gettier cases as thought experiments is unorthodox in that most analyses take them to bear upon our concept of knowledge, not about knowledge itself, metaphysically speaking. After all, is this not what the differences between Occidental and Eastern subjects alleged by Stich and his associates purport to show? They diverge on their concept of knowledge, or if one prefers, on the meaning of "knowledge," since the former takes it to mean implicitly "justified non-accidental true belief," whereas the latter seem ready to accept, at least

---

[22] Wittgenstein explicitly defended this view when he said, what Mach takes to be a thought experiment is a mere "grammatical consideration," see Wittgenstein, *Nachlass*, 107, 284–285, quoted in Bouveresse, *La connaissance de l'écrivain*, 116.

in some cases that justified accidental true beliefs can be knowledge. According to Stich, there is no essence of knowledge which could lie behind our intuitions, and the very fact that they conflict shows that no such essence is to be found. On the contrary, the traditional epistemologist claims that the Gettier counterexamples show that justified true belief is not a necessary and sufficient condition on our concept of knowledge, hence that we have to look for other conditions.

Williamson strongly disagrees with this. He denies that there are conceptual truths and that philosophy is in charge of articulating them. His analysis of thought experiments is in part motivated by his criticism of the epistemological conception of analyticity, and of the classical view that philosophy is an a priori discipline. I cannot deal here with his conception of philosophy.

What would an analysis of the Gettier thought experiment involve along conceptualist lines? It would imply that (1) above is a conceptual necessity about our concept of knowledge, and that (2) above is a conceptual possibility. But, objects Williamson,

> The conclusion would be that it is conceptually possible to have justified true belief without knowledge. That does not refute the hypothesis that knowledge just is justified true belief, of metaphysical necessity, any more than the conceptual possibility of something with atomic number 79 that is not gold refutes the hypothesis that gold just is the element with atomic number 79, of metaphysical necessity. The primary concern of epistemology is with the nature of knowledge, not with the nature of the concept of knowledge. If knowledge was in fact identical with justified true belief that would be what mattered epistemologically, irrespective of the conceptual possibility of their non-identity.[23]

Moreover, Williamson argues, the conceptualist reading would be trivial, for on any reasonable understanding of "conceptually possible" it is conceptually possible that some abnormal instance of the Gettier case is not a case of justified true belief.[24] In other words, we would always cook up an abnormal situation, which would show that the concept of knowledge and that of justified true belief come apart.

Williamson's argument presupposes the view that knowledge is, like gold or water, a natural kind.[25] But it is not evident that the concept of knowledge is unified and has a real essence in the sense in which natural

---

[23] Williamson, *Philosophy of Philosophy*, 205.

[24] Ibid., 205.

[25] This is the thesis defended by Kornblith; see the exchange between Kornblith, "Timothy Williamson's" and Williamson, "Replies."

kinds are supposed to have a real essence. There are borderline cases of knowledge, such as the one described in the situation invented by Lehrer of Mr. Truetemp:

> Suppose a person, whom we shall name Mr. Truetemp, undergoes brain surgery by an experimental surgeon who invents a small device which is both a very accurate thermometer and a computational device capable of generating thoughts. The device, call it a tempucomp, is implanted in Truetemp's head so that the very tip of the device, no larger than the head of a pin, sits unnoticed on his scalp and acts as a sensor to transmit information about the temperature to the computational system of his brain. This device, in turn, sends a message to his brain causing him to think of the temperature recorded by the external sensor. Assume that the tempucomp is very reliable, and so his thoughts are correct temperature thoughts. All told, this is a reliable belief-forming process. Now imagine, finally, that he has no idea that the tempucomp has been inserted in his brain, is only slightly puzzled about why he thinks so obsessively about the temperature, but never checks a thermometer to determine whether these thoughts about the temperature are correct. He accepts them unreflectively, another effect of the tempucomp. Thus, he thinks and accepts that the temperature is 104 degrees. It is. Does he know that it is?[26]

A knowledge reliabilist, for whom knowledge is but reliable true belief, will accept Mr. Truetemp's case as a genuine case of knowledge, even though the situation is, admittedly, a bit abnormal. But an internalist about knowledge will disagree. However, it is hard to see how this abnormal case, which is admittedly conceptually possible, would be trivial, for it is precisely meant to elicit in us our intuitions about two different concepts of knowledge, the reliabilist and the externalist one. This is, after all, what thought experiments in philosophy are about. Indeed, one can agree with Williamson that if Mr. Truetemp's case were impossible, this would indeed show something about the *nature* of knowledge, not merely about our concept of it. But this is precisely what is in question! There is a conflict—and a genuine one, as such thought experiments show—between two concepts of knowledge. The thought experiment is meant to illustrate the conflict.

To return to the Gettier case, we can grant Williamson's claim that Gettier cases bear upon *knowledge*, not upon our concept of knowledge. But that does not imply that we cannot have diverging conceptions of what knowledge is. Internalist versus externalist notions of knowledge,

---

[26] Lehrer, *Theory of Knowledge*.

coherentist versus foundationalist notions, or virtue theoretic versus reliabilist notions are familiar oppositions within the philosophy literature on knowledge. When philosophers in the course of an argument use thought experiments in favour of one or other of these notions, they do not pretend to elicit the true concept just by relying on the intuitions, be they Asian, European, male, female, Catholic, or Buddhist. They intend to *argue* in favour of one conception or another, and thought experiments are part of their argument. But they are not their *whole* argument. This is where both the rationalist *a priorist* and the experimental philosopher are wrong. They expect to draw *the* concept from a thought experiment, and if they do not find it they conclude, like the eliminativist experimental philosopher, that there is no concept at all, hence (depending on the issue) no knowledge, no persons, no freedom etc. But this is wrong. For there are philosophical conflicts that can only (or perhaps cannot) be solved by an argument. Thought experiments are not reports about our "folk" concepts.

Thought experiments can indeed be used in order to analyse folk concepts and our common sense scheme. But even so, this does not imply that it is their only use. They are used in philosophy in the context of offering hypotheses, for the sake of arguments about real things, and not simply about our concepts of them. Thought experiments are artefacts within arguments, which purport to take us from true premises to true conclusions. In this respect, thought experiments are only in part the reports of discoveries about our beliefs or about our concepts. They are constructed for the sake of an argument. They are hypotheses that we put forward to see what follows. In some cases the deliverances are convincing. In other cases they are not, because there is a conflict between various concepts. The conflict may or may not exist in ordinary thought.

It does not follow that philosophical thought experiments can be put on a par with scientific thought experiments. Discussing the psychological experiments adduced by Stich and others, Williamson says:

> Native English speakers sometimes dispute the Gettier verdict ... In doing so, they show poor epistemological judgment but not linguistic incompetence: they are not usually accused of failing to understand the relevant words of English; it would be inappropriate to send them off to language school for retraining.[27]

---

[27] Williamson, *Philosophy of Philosophy*, 188.

I agree. But that does not show that a well-devised thought experiment, constructed in the course of an argument about knowledge, can elicit diverging conceptions. After all, Putnam's Twin Earth case elicits diverging conceptions of the nature of mental content or of semantic reference. If we had resolved the issue in favour of externalism contra internalism, or in favour of one conception or another on knowledge, we could certainly say that the Gettier case, or the Mr Truetemp case, has offered genuine counter-possibilities, or not. But unless these conflicts are resolved, the question is open. In most philosophical cases it is what happens.

Wouldn't it be nice if philosophy could, like science, reveal essences? Indeed, I agree with Williamson that this is what it ought to *aim* at. I do not accept the Wittgensteinian view[28] that philosophy is *entirely* conceptual. But it has to be, when a question is not solved. And looking at thought experiments could help in setting the problem, sometimes in solving it. So it does not seem to me that Williamson has shown that all thought experiments deal with metaphysical, and not conceptual possibilities, although I agree that in philosophy we have both.

7. *Conclusion*

So the scope of the counterfactuality thesis seems to be rather limited. It actually does not tell us how we can in general assess a thought experiment, or evaluate the claims that it makes. It does not give us a way to decide which of the main conceptions (i)–(iv) above of thought experiments are correct. But this is hardly surprising. If the counterfactuality thesis is correct, there are as many different kinds of thought experiments as kinds of counterfactual reasonings. The counterfactuality claim is not the thesis that we can illuminate the nature of specific thought experiments by translating them into counterfactual claims. It is just the thesis that philosophical experiments do not use resources beyond those of counterfactual reasoning. And given that counterfactual reasoning is a species of common sense reasoning, this rules out two antagonistic views of thought experiments: the ultra rationalist one, which says that thought experiments appeal to a power of *a priori* intuition of essence, and the ultra-empiricist one, which says that they appeal to empirical beliefs only.

[28] Reinstated angrily by Hacker in "A Philosopher of Philosophy" against Williamson.

If this is correct, philosophical thinking, insofar as it uses characteristic thought experiments is neither conceptual nor empirical. It can be both: sometimes conceptual, sometimes not.[29]

[29] Versions of this article have been read in the Athens Conference on Thought Experiments in April 2007, and in the Grenoble conference in June 2008, which was its sequel. I would very much like to thank Sophie Roux and Katerina Ierodiakonou for their invitation, their discussions and their patience in editing this text; Jean-Yves Goffi for his discussion of this article in Grenoble; Stelios Virvidakis, Christophe Grellard, Carla Rita Palmerino, and all the participants in these conferences for their remarks. In January 2009, my discussion of Williamson's book at the VAT conference in Tilburg helped me to revise in part my initial views. I thank Filip Buekens, Igor Douven, and the participants for the invitation and their feedback. And thanks to Timothy Williamson for having clarified for me some points discussed in paragraph 5 above.

ON THE VERY IDEA OF A THOUGHT EXPERIMENT

Jean-Yves Goffi and Sophie Roux*

Even after the characterisation of thought experiments suggested in the introduction of this book, their population—we are a bit tempted to speak of a bestiary or zoo full of monsters—remains too great for us not to wish to tidy it up a bit: in other words, to distinguish among different types of thought experiment. Several attempts have been made to do so, and we dispose of various taxonomies. You can distinguish thought experiments by the disciplines to which the matter in question belongs, or by their function in a theory, descriptive or evaluative thought experiments, destructive and constructive thought experiments, thought experiments that result in decidable cases and thought experiments that do not, thought experiments that correspond to a physically possible experiment and thought experiments that correspond only to a metaphysical possibility, thought experiments that justify a law and thought experiments that clarify a concept, and so on.[1] Nonetheless, all these taxonomies miss what will prove to be not only a useful distinction to tidy up our zoo, but a response to the radical objection that could be applied to the very idea of a thought experiment. If imagination is free to wander about in a thought experiment, without anything equivalent to the principle of reality able to intervene, is it ever possible that a thought experiment would not work *per se*? Of these two options, only one is real:

- Either all thought experiments are successful, and there is no need to be a stubborn falsificationist to conclude that the very idea of a thought experiment should be abandoned.
- Or there is a distinction between successful and unsuccessful thought experiments, and the idea of a thought experiment could have a certain epistemological depth.

---

* We thank Marc Naimark for having revised this text, and John Zeimbekis for having commented on its first version.

[1] For examples of taxonomies, see Popper, *Logic*, 442–452; Brown, *Laboratory*, 33–48; Sorensen, *Thought Experiments*, 197–201; Rescher, *What if?*, 47.

In that sense, showing that there is a distinction to be made between successful thought experiments and unsuccessful thought experiments and explaining why there is such a distinction aims not only at making a distinction for the sake of "taxonomophilia." The point is also to show that the idea of a thought experiment is a substantive idea that has a certain epistemological depth.

Distinctions between legitimate and illegitimate uses of thought experiments could already be encountered in earlier philosophy of science.[2] But our question about the conditions for successful thought experiments has rarely been addressed in recent literature, in particular if it is seen as a way of defending the very idea of thought experiments. To answer this question, we will proceed in three steps. First, we will expose the problem and present our general method to answer it. Secondly, we will analyse some examples of famous successful thought experiments in order to formulate a hypothesis on the conditions of a successful thought experiment. Thirdly, by offering examples of unsuccessful thought experiments we will illustrate what happens when one of these conditions is not fulfilled. In conclusion, we will make a few remarks on the whys and wherefores of our argumentation.

1. *The Problem and the Method*

Let us be granted our impression that there are thought experiments that work. Inasmuch as thought experiments are arguments, "to work" here means, at least to begin with, nothing other than to be successful as an argument, in other words to be able to convince one's interlocutors. A first *caveat* is necessary here: writing that thought experiments are arguments might appear as an unjustified bias against Tamar Szabó Gendler, who recently argued against John Norton that at least some thought experiments are not justifiable on the basis of straightforward arguments.[3] It might be an unjustified bias, but in this paper we do not need the strong Nortonian thesis that thought experiments are *only* arguments: we will be satisfied with the thesis that they are arguments, a thesis that even Gendler would not refuse. Indeed our question is simply the following: inasmuch as thought experiments are arguments, and notwithstanding

---

[2] For example, Duhem, *La Théorie*, 306–309; Popper, *Logic*, 442–456.

[3] Gendler, "Galileo"; Gendler, "Thought Experiments Rethought."

the fact that some of them might involve a contemplation of an imaginary scenario, how is it that some of them work, while others do not?

In one of his papers, Norton asks precisely the question about the mark that could identify successful thought experiments; he ends up writing that "the mark is just that the thought experiment either uses an argument form licensed by a logic or can be reconstructed as one."[4] But, since every argument must respect standard logic, this cannot be a sufficient mark for identifying successful thought experiments: they have to satisfy more precise conditions. Nor would a reference to truth help here: it may happen that shared assumptions between the interlocutors yield false conclusions. Note in particular that a thought experiment can be successful in the sense that a majority of people admit that it has a certain outcome that is considered as intuitively obvious, although later experiences or further theorising will prove that it is not correct: this is for example the case of the probability paradoxes, well-known to every college freshman. In the conclusion of this paper, we will come back to the issue of relativism that our perspective may raise; to begin with, we will, however, not consider this issue, but focus instead on the question of how one can be convinced by some thought experiments and not by some others.

We find in the secondary literature various hypotheses to account for the feeling that there is a difference between thought experiments that are convincing while others are not. According to a first hypothesis, in the case of thought experiments that work, laws of nature or even natural kinds are preserved.[5] According to a second hypothesis, our intuitions ground thought experiments that work.[6]

However, we feel that the question as to whether laws of nature are preserved is not pertinent. If the thought experiment in question takes place in the realm of the natural sciences, then it would be, at the least, unsatisfactory to speak of preserving the laws of nature without further

[4] Norton, "Why Thought Experiments," 53–54. Our departure from the question of the nature of thought experiments to the question of the conditions they respect when they are successful is also to be found in Janis, "Thought Experiments Fail," 113–118; as for Atkinson and Peijnenburg, "Thought Experiments Poor," they show that unsuccessful thought experiments, whether in science or in philosophy, share the two same features: they lead to contradictory conclusions or to conclusions that beg the question.

[5] This is a point made by Wilkes, *Real People*, 12–15, *passim*, against thought experiments concerning personal identity, most notably Parfit's people splitting like amoebas. See also Jamieson, "Method and Moral Theory," 484–486; Brown, *Laboratory*, 30–31.

[6] Brown, *Laboratory*; Gendler, "Galileo," and Gendler, "Thought Experiments Rethought," albeit with completely different conceptions of intuition.

specification, for what one is precisely seeking to establish by means of this thought experiment might be itself a law of nature. If the thought experiment in question does not concern the natural sciences, then the laws of nature could be preserved, and yet no intuition may be at hand, because what is at stake is not a law of nature.

The second hypothesis—that thought experiments that do not work leave us lost in terms of intuition—does indeed correspond to the feeling one can experience faced with certain thought experiments. We would like, however, to know a bit more about this feeling and what can be understood by "intuition." Certainly, speaking about intuition does not necessarily imply that we attribute to people a faculty of direct insight into supersensory truths: experience of things and people indeed yield some intuitions as to what two equally heavy bodies are, how most people would behave in a given situation, when a body is said to be faster than another, what it is to be a good person or to be a bad person, etc. But, even granting this, intuition is a notoriously ambiguous notion. Moreover, it would remain to determine what an intuition that intervenes in a thought experiment is, and how it could be not only psychologically persuasive, but convincing in a controversial context.

This is what we will describe in this paper, using a method inspired by a counterfactual thought experiment presented by Nicholas Rescher.[7] We use his example because of its great pedagogical value, but it will soon be clear that its strong counterfactual dimension is not what is most significant about it. Let us begin with the principle that a counterfactual thought experiment is in reality counterfactual reasoning—to be distinguished from counterfactual declarations, which pose totally different problems and raise totally different questions. In counterfactual reasoning, one introduces a supposition held to be false: "Caesar never crossed the Rubicon"; "General de Gaulle was killed in 1940 during the battle of France." This is the beginning of a "what if" scenario. What follows is the development of the consequences of this supposition: "the Roman Republic is not overthrown," "the Resistance is dominated by the Communists." Rescher wondered just what happens in these cases. The principle of his response is as follows: the counterfactual arguments aim to determine the consequences of the introduction of a supposition that contradicts our beliefs, that is to say, that is in conflict with all we assume to be true, at least in one region of our knowledge, a region as broad or

[7] Rescher, *What if?*, 98–99.

narrow as we wish. The difficulty is that, if these initial beliefs are consistent, we cannot preserve them all under the supposition that we have replaced one of them by a new belief that is the negation of the first. The issue is thus, in Rescher's words, to "break the chain of inconsistency at its weakest link." Let us reason, following Rescher, about a well-known example and suppose that we introduce the following counterfactual: "If this rubber band were made of copper ..."

Well, if this rubber band were made of copper, what then? There are two possibilities, the first one "natural":

(a) If this rubber band were made of copper, then it would conduct electricity (since copper conducts electricity);
the second one "unnatural":
(b) If this rubber band were made of copper, then it would not conduct electricity (since this band does not conduct electricity).

Why do we consider (a) a more natural conclusion? We can explain this by making explicit part of the beliefs of the "experimenter."

(1) This band is made of rubber.
(2) This band is not made of copper.
(3) This band does not conduct electricity.
(4) Things made of rubber do not conduct electricity.
(5) Things made of copper do conduct electricity.

(1)–(5) are a consistent set of beliefs in the textbook sense: it is possible for all the statements expressing them to be true. If we introduce as a counterfactual supposition the negation of (2), i.e. non-(2), we must immediately reject (2) by definition, as well as (1)—for the reason that the same object cannot be both copper and rubber (at the same moment and in the same proportions). Non-(2) was introduced at the price of evicting (1) and (2): what else can be kept or rejected? We can keep (4) which expresses a general law relative to the conductivity of rubber, and whose value of truth is not affected by the choice of non-(2) rather than (2). We can still choose to keep (3) since it states a true fact relative to this rubber band (it does not conduct electricity). But we would then have to reject (5) because non-(2), (3) and (5) cannot all be true at the same time. The choice of keeping (3) thus leads us to the "non-natural" conclusion. The "natural" conclusion will therefore be interpreted in the following fashion: by introducing non-(2) we reject (2) and (1). We can keep (4) for the same reasons as before. And this time, we choose to keep (5), since it states a true fact relative to things made of copper. But if we decide to keep (5), we have to eliminate (3) because non-(2), (3) and (5) cannot all be true at the same time.

We have employed expressions such as "we decide to keep" or "we choose to preserve." This gives the impression that the process, at a critical point, is fully arbitrary. The issue is not with the initial supposition: that is part of the normal regime for thought experiments, because of the counterfactuality clause. No, the problem lies rather in the way the procedure plays out to determine the weakest link; why reject (3) rather than (5)? Rescher has a general answer to this objection: it's a matter of cost-benefit optimisation relative to one's overall commitments. In the case of the rubber band made of copper, the optimisation is obvious: Rescher remarks that it would cost more to dismiss general statements about copper things like (4) and (5) than particular statements such as (1), (2) and (3). It is therefore not arbitrary to seek to preserve the former, and to find the weak link among the latter.

The lesson we intend to learn from this example is of general order. We will summarise it in three points:

i. Roy A. Sorensen and Soren Häggqvist resort to modal logic to differentiate different types of thought experiments.[8] It is indeed legitimate, given the equivalence of counterfactuals and modalities established by Stalnaker.[9] But we are not interested in this; rather, we aim at understanding what pragmatic conditions successful thought experiments satisfy. For that, modal logic is not necessary, since we succeed in dispensing with it.
ii. Thought experiments are procedures that resemble simple tests of consistency. However, they not only reveal that the whole of the experimenter's beliefs have become inconsistent, they also orient him towards an alternative in which each of the branches has an acceptance/rejection basket.
iii. Thought experiments invite the experimenter to seek the weakest link in her body of beliefs. It is in the ordering of the beliefs, not necessarily perceived before the thought experiment was carried out, that the thought experiment will make the weakest link obvious.

---

[8] Sorensen, *Thought Experiments*, 133–166, and Häggqvist, *Thought Experiments*, 92 sqq. For a shorter presentation of Sorensen's theses, see Sorensen, "Thought Experiments and Epistemology."

[9] For a short reminder on this question, see Engel's essay in this volume.

In this sense, a thought experiment is a test whose goal it is to reconfigure the organisation of our beliefs. In that, it is no different from a real experiment. But the confrontation with physical reality is not necessary: the reconfiguration can happen even following a counterfactual scenario.

2. *Successful Thought Experiments*

Equipped with the basics provided by Rescher's example, we can enter the zoo of thought experiments and try to understand the reasons why some of them are successful. We will now visit an extremely familiar animal, a Galilean pet; we will then climb on the shoulders of Ernst Mach, albeit to advance theses distinct from his; lastly, we will seek confirmation of our theses by heading to the other side of the zoo to visit an unconscious violinist.

2.1. *A Familiar Animal: Galileo's Two Bodies Strapped Together*[10]

The domain of this thought experiment is physics. It not only reveals the inconsistency of a certain set of beliefs, it orients the experimenter towards a new belief. This is why it has been described as being simultaneously destructive and constructive:

- destructive aspect: to refute the Aristotelian proposition (a)
  (a) two bodies of the same material descend in the same medium with speeds proportional to their weights;
- constructive aspect: to replace it by the Galilean proposition (b)
  (b) two bodies of the same material descend in the same medium with like speeds.[11]

What happens if, in a world where (a) is a law of nature, we take a heavy body A and a light body B, strap them together and drop them from a certain height? Our presentation of the thought experiment follows step

---

[10] This thought experiment was already presented in the *De motu*, but we will refer here to the *Discorsi*, in Galileo, *Two New Sciences*, 107–109. As Palmerino's essay in this volume reminds us, it is probably the most widely commented thought experiment in literature. See for example Popper, *Logic*, 442–443; Koyré, "*De Motu gravium*," 224–232; Brown, *Laboratory*, 77–79; idem, "Thought Experiments," 122–126; Sorensen, *Thought Experiments*, 126–127; Norton, "Are Thought Experiments," 340–345; Gendler, "Galileo," 397–424; Atkinson "Experiments"; Kühne, *Gedankenexperiment*, 31–57, *passim*.

[11] Brown, *Laboratory*, 77–79, and idem, "Thought Experiments," 124, called "Platonic" the thought experiments that are simultaneously constructive and destructive; in so far as they are constructive, they would yield an *a priori* knowledge of nature.

by step the dialogue between Salviati and Simplicio. Note that the bone of contention between recent commentators does not concern its logical reconstruction, which is fairly consensual, but the nature and function of the assumptions necessary to draw a conclusion.

(1), from (a): the heavy body A and the light body B fall with speeds $v_B < v_A$; Simplicio accepts the assumption (A_1): natural speed is mediative, i.e. that if a slower body is connected to a faster body, then the slower will slow down the faster and the faster will speed the slower; hence (2), from (1) and (A_1): $v_{A+B} < v_A$.

But a broad and sweeping assumption, here formulated, is (A_2): weight is additive, i.e. that the weight of a body combined of two bodies is the sum of the weights of these two bodies, $W_{A+B} = W_A+W_B$; hence (3), from (A_2): $W_{A+B} > W_A$; and finally (4), from (3) and (a): $v_{A+B} > v_A$.

From the contradiction of (2) and (4), we infer non-(a): two bodies of the same material do not descend with speeds proportional to their weights. We consequently give up (a) and substitute for it (b): two bodies of the same material descend in the same medium with like speeds. In order to make this last step, we need a last assumption (A_3), that weight is the only parameter to take into account when we study falling bodies; the warrant for (A_3) is the fact that the initial Aristotelian proposition (a) concerns bodies from the same material falling in the same medium.[12]

From a logical point of view, this thought experiment is a *reductio ad absurdum* of (a). This *reductio ad absurdum* works if, and only if, we admit assumptions (A_1), (A_2) and (A_3), which is not obvious from a physical point of view, the existence of alternative physical scenarios being enough to destroy the Brownian view that this would be a "Platonic" thought experiment that gives us the occasion to acquire an *a priori* knowledge of laws of nature. For example, against assumption (A_1), that natural speed is mediative, we could say that (a) concerns only *natural* bodies that have *natural* speeds.[13] But strapping (or tying or gluing) two bodies together does not make one *natural* body and that, consequently, (A_1) simply does not apply—we may know a law concerning the fall of natural bodies, not a law concerning the fall of artificially attached bodies.[14] Or, against assumption (A_3), that weight is the parameter to take into account in the

[12] Galileo, *Two New Sciences*, 105–106. Brown, *Laboratory*, 78, and idem, "Thought Experiments," 125, claim that the Galilean theory is not logically deduced from known data precisely because he does not take this premise into account.

[13] Galileo, *Two New Sciences*, 105–106, makes clear that speed is determined by nature.

[14] This objection is raised by Koyré, "*De Motu gravium*," 231–232. See also Gendler, "Galileo," 405–406, who objects moreover that an Aristotelian would introduce a

case of falling bodies, we could say that we have to take into account the configuration of bodies, as well (which is trivially true: a ball and a plank of wood do not fall the same way)—or variations of the gravitational field (which is true if you are a 20th century physicist)—or colours and odours of bodies (which is not true according to Aristotelian, Galilean or Einsteinian physics).[15]

Our point here is of course not to charge Galileo with logical weakness or with physical incompetence. Nor is it to determine if an Aristotelian of the 17th century would have shared the view Galileo ascribes to his Aristotelian Simplicio. Rather, it is to ask the question as to why the Galilean Salviati can plausibly present this argument as a "short and conclusive argument" "even without further experiment" to the Aristotelian Simplicio.[16] According to us, if Salviati can present his argument as conclusive, it is because:

i. (A_1) (A_2) (A_3) are assumptions that both the Aristotelian and the Galilean of the dialogue agreed upon.
ii. Both of them consider that it would cost more to give up one of these basic assumptions than to give up (a).

Can we go farther and explain why Simplicio and Salviati agree on certain assumptions and judge that these assumptions are stronger propositions than (a)? In the case of Rescher's rubber band, the priority between beliefs was easy to determine because it amounts to the priority of one general statement (copper conducts electricity) over a particular statement (this band does not conduct electricity). In the case of Galileo's strapped bodies, the process of prioritisation is not that obvious: for example, (A_1) could have been taken as a premise in the construction of a concept of speed.

2.2. *Some Zookeepers: Mach and Beyond*

To the question we just asked, Mach gave at first sight an interesting answer, namely, that the two protagonists of the *Discorsi* have stores of instinctive knowledge that comes from our former experiences with

distinction between united and unified bodies, or say that the speed of fall depends on the degree of connectedness between the strapped bodies.

[15] Atkinson, "Experiments," 221–222, considers a world in which there are variations of the gravitational field; Atkinson and Peijnenburg, "Galileo," 128–133, examines moreover situations of falling in resisting or turbulent media.

[16] Galileo, *Two New Sciences*, 107.

bodies. In his masterful work, *The Science of Mechanics*, Mach asked an important question: how is it that the result of a thought experiment may have more authority than the result of a real experiment? The answer that, as a good empiricist, he gave to this question is that a thought experiment taps into a store of instinctive knowledge and summons up beliefs from this store with respect to a specific problem. As explained in the introduction of this book, Mach's notion of a thought experiment is however a complex one, that needs some clarification before being used. This has not prevented the vulgate that claims affiliation with Mach to say that there are intuitions that intervene in thought experiments and to attribute to them in a very general fashion the following characteristics:

i. Not only are they tacit beliefs but, insofar as they result directly from the silent impression produced by nature in the space of our representations, they are barely even propositional attitudes.
ii. Taken together, they constitute a shapeless whole, contrary to the organised structure that is that of constituted science.
iii. They are primarily common intuitions coming directly from our innate knowledge or from basic experiences with bodies.[17]

But, if the intuitions that intervene in thought experiments had these three characteristics, it would be impossible to understand the mechanism of thought experiments. Better to say that:

i. A thought experiment does not have as its material any representation that would be the direct impressions that nature leaves in us, but rather, beliefs. It is usually said that beliefs intervening in thought experiments are "implicit" or "tacit," but they are not implicit in the sense that they would constitute a kind of mysterious knowledge. They are implicit simply in the sense that they are unexpressed, while being "close at hand," to use the Machian expression: they may be there, just under the surface, the function of the thought experiment being to make them come to the fore.[18] For example, in the Galilean thought experiment, assump-

[17] Gendler, "Galileo," 414–415, and idem, "Thought Experiments Rethought," 1161–1162. In the conclusion, we explain more thoroughly the reasons why we cannot agree with Gendler.

[18] John Zeimbekis objected that the relevant belief for a thought experiment I am performing may not be already there in my set of beliefs: I don't after all have infinite beliefs, i.e. I don't have all belief—or disbelief—states corresponding to *all* possible well-

tions (A_1), (A_2) and (A_3) are of the same nature, even though (A_2) is not expressed: it would have been if, contrary to what our thousand-year-old experiences with balances and other simple devices show, the 17th century Aristotelians had defended the thesis that weights are not additive.

ii. Our store of intuitive knowledge is not shapeless, but has a hierarchical organisation. By this, we mean that there are links between beliefs and that there is a distinction between stronger and weaker beliefs. The evidence for this hierarchical organisation is that when we abandon one belief, other beliefs fall with it, and that we at times prefer to give up certain beliefs so as to be able to keep others. Retrospectively, we say that the beliefs we gave up were "weaker" than the "stronger" ones we kept, or, as we did after Rescher, employ the metaphor of cost-benefit optimisation.

iii. You must accept all kinds of beliefs in these stores of knowledge, and not only beliefs inspired by a passive observation of nature or by an elementary manipulation of bodies. To confine ourselves to the domain of physics, about which Mach was speaking, the latest results of more formalised sciences may intervene as beliefs just like the know-how of engineers. In other words, the distinction between "commonsensical" beliefs and "sophisticated" beliefs is not a relevant distinction here; the only relevant distinction is that between what is accepted as common ground by performers of the thought experiment, and what is not. This is why Galileo makes explicit that the Aristotelian Simplicio agrees to Salviati's assumptions. If the extension to "sophisticated" beliefs is granted, as we think it should be, it can be understood why there are thought experiments not only at the more primitive stages in the history of mechanics, but at other stages as well.

In this respect, it is interesting to come back to the example that initiated Mach's reflections on thought experiments in the *Science of Mechanics*, Stevin's necklace. Remember that the crucial assumption for this thought experiment is the impossibility of perpetual motion: if this impossibility were not granted, there would be nothing to work on, the other assumption (when one removes equal things from equal things, one has

formable propositions from the concepts I possess. True enough: we cannot decide if thought experiments help beliefs to crystallise or simply to come to the fore. But we will let the question undecided, because a justified decision would lead to a discussion on tacit knowledge that we cannot afford here.

equal things) being untouchable. Not surprisingly, Mach claimed that it is something known instinctively:

> Unquestionably in the assumption from which Stevin starts, that the endless chain does not move, there is contained primarily only a *purely instinctive* cognition. He feels at once, and we with him, that we have never observed anything like a motion of the kind referred to, that a thing of such a character does not exist. This conviction has so much logical cogency that we accept the conclusion drawn from it ... without the thought of an objection, although the law, if presented as the simple result of experiment, or otherwise put, would appear dubious.[19]

Mach may lead the casual readers astray here. The impossibility of mechanical perpetual motion would not be granted by everybody—not only insomuch as there will always be a nut who will believe in something aberrant, but because this principle has a history. This history remains to be written, but its main lines are clear: during the late Middle Ages, many projects of perpetual motion were elaborated; during the 16th and 17th centuries, engineers and scientists began to treat the impossibility of mechanical perpetual motion as an assumption needed in physical proofs; this assumption was finally the object of a demonstration in Lazare Carnot's *Essai sur les machines en général* (1783). The belief that mechanical perpetual motion is impossible, that Mach treated as an immemorial instinctive knowledge common to all men, thus emerged slowly in history and was at the beginning of the 17th century a rather sophisticated belief, admitted only by engineers and scientists.

Hence, the outcome of our jaunt with Mach is twofold. First, the Machian notion of instinctive knowledge is not clear enough to be a real explanation of why certain processes of prioritisation happen rather than others; better to leave it unexplained. Secondly, what Mach called "instinctive knowledge" and that his followers call "intuitions" may be more aptly described as beliefs, in order both to avoid the mysticism surrounding the notion of intuition and to include more or less elaborated kinds of statements. This might be enough to suggest that a thought experiment works not because it operates in one field or another, but rather because is has as its material certain beliefs, beliefs that are shared and that constitute a hierarchically organised whole. Before coming to a fuller presentation of our hypothesis, we will test it by leaving the

[19] Mach, *Science of Mechanics*, 34. Following in Mach's footsteps, Gendler, "Thought Experiments Rethought," 1161–1162, claims that this thought experiment "evokes certain quasi-sensory intuitions" and that the resulting belief is produced "not inferentially, but quasi-observationally."

menagerie of falling bodies of the 17th century to consider a famous 20th century freak, Judith Jarvis Thomson's unconscious violinist.

2.3. *The Exotic Aviary: Thomson's Violinist*[20]

The domain of this thought experiment is applied ethics. It is an argument against opponents to abortion, the outcome of which is to distinguish between two senses of the expression "right to life."

Thomson begins by a reconstruction of her opponents' argument: Point (a) grants that the foetus is a person from the moment of conception, *i.e.* she gives her opponents' thesis its greatest strength. Point (b) consists in a kind of syllogism: every person has a right to life; but a foetus is a person; hence a foetus has a right to life. Point (c) is a bit more difficult. It begins with a set of premises, from which a conclusion is drawn.

Two broad and sweeping assumptions are necessary to begin with:

(A_1) There are differences between the stringency and strength of rights.
(A_2) It is morally wrong to give more weight to a right whose stringency and strength are weak than to a right whose stringency and strength are great.

Thomson's opponents' argument is the following:

(1) A woman has a right to decide what shall happen in and to her body (right to autonomy).
(2) A person's right to life is stronger and more stringent than another person's right to autonomy.
(3) From point (b): A foetus is a person.
(4) An abortion (the fact of voluntarily killing a foetus) performed in the name of autonomy is a case of giving more weight to a right whose stringency and strength are weak than to a right whose stringency and strength are great.

Hence, they conclude:

(C_1) Abortion is morally wrong (Evaluative conclusion).
(C_2) An abortion may not be performed (Practical conclusion; another broad and sweeping assumption is at work here: (A_3) Morally wrong acts may not be performed).

---

[20] Thomson, "Defence of Abortion."

Thomson's strategy consists in performing a thought experiment showing that (2) is the weak link in this set of premises; more precisely, showing that, if there is an equivocation in (2), then the argument is not sound.

Suppose you wake up one morning and find yourself back-to-back in bed with an unconscious violinist. He suffers from a fatal kidney ailment and the Society of Music Lovers has discovered that you alone have the right blood type to help. They have therefore kidnapped you and plugged his circulatory system into yours, so that your kidneys can be used to extract toxins from his blood as well as yours. But the Director of the Hospital is very comforting: "Put your mind at ease, it's only for nine months."[21]

This is a clear case of treating a person as a means only, not only an intuition that the members of the Society of Music Lovers are behaving strangely or rudely. And it is premise (2) that falls under attack: broad and sweeping assumptions (A_1) and (A_2) are common to both Thomson and her opponents, as is premise (1).[22] Premise (3) has to be agreed on since it is the conclusion of a sound argument. True, premise (4) is not accepted, but only because (2) has been rejected to begin with, and as a consequence of this rejection.

One can rationally refuse to admit (2) because (2) mixes up two senses of "right to life." "The right to life" can mean either "the right not to be arbitrarily killed," or "the right to receive full support to life." Consequently, (2) may mean:

- either: (2^a) a person's right not to be arbitrarily killed is stronger and more stringent than another person's right to autonomy (which seems reasonable enough as it boils down to saying that I may not wantonly kill someone as an expression of my personal autonomy: actually, it is the core idea in the good old Lockean theory of the right to life).

[21] The initial clause, "suppose you wake up one morning and find yourself back to back in bed with an unconscious violinist" is a rhetorical artifice: in order to understand this thought experiment, there is no need for the reader to take on the situation and the reactions of the person who would wake up like this one morning. This differs from the case of Jim and the Indians recalled in the introduction of this book, where the reader must, in a certain fashion, become Jim to understand what is in reality the deep indifference supposed of Utilitarians with respect to the question of personal integrity, and which raises issues for them.

[22] The scenario says: "No doubt, the mother has a right to decide what shall happen in and to her body; everyone would grant that."

- or: (2^b) A person's right to receive full support to life is stronger and more stringent than another person's right to autonomy (which may be true under certain circumstances, but which is not true without serious qualifications, so that the burden of the proof is still incumbent to the members of the Society of Music Lovers).

A few qualifications may be added to this logical presentation. First, one could wonder how it is possible that an unconscious violinist desperately needs another's body for life support: here, it is the possibility of the scenario—that is to say its credibility—that is in question. The response can be relatively direct: an unconscious violinist may desperately need another's body in the sense that the victim of a car accident may have a desperate need for a transfusion with a rare blood type. In this case, urgency may push doctors to very strongly insist that any staff member with this rare blood type accept to donate their blood: Thomson proceeds by passing to the limit of these well-known cases, and this seems a respectable procedure.

More seriously, it may be noted that Thomson's thought experiment relies on an analogy between a foetus and an unconscious violinist. The obvious objection is that this analogy is not obvious: the relation of the foetus to the pregnant woman is not of the same nature as the relation of the comatose violinist to a perfect stranger. But in the same way that in real experiments we do not take into account all that is actually encountered, likewise a thought experiment does not have to take everything into account. At issue is the extension of the concept of "the right to life." The very "logic" of the language of rights is to clearly distinguish between virtual ethical and legal rights on one hand, and real psychological and affective relations on the other. One can, of course, deplore this. But that does not go against Thomson's thought experiment, since the idiom of rights is hers, but also that of her adversaries.

Lastly, we can wonder what would happen if we were dealing with someone who believed that it is proper, given these imaginary circumstances, to authorise the Society of Music Lovers to make use of the body of another, against his will, for at least nine months, so as to permit the comatose violinist to continue to play, as her exceptional musical talent is a divine gift. There is no point in stating here that such a conviction would be excessive, for the response is only too predictable: "it is not because an opinion is held by a minority that it is excessive." But the answer can be as follows: Anyone who would speak in this way would put herself out of bounds, so to speak, with respect to the language of rights. She would adopt another argumentative posture, admitting, for example, that life is a gift of God, and that human beings must act in consequence. But beyond the fact that such an affirmation has its own well-known weaknesses,[23] it is not pertinent with respect to everything that had been admitted by the interlocutors when the agreement on the premises was slowly but surely established.

Thomson's thought experiment reveals that, just as in public or executed experiments, thought experimenters need to share beliefs and to hierarchise beliefs if they want to reach a conclusion: in this respect, a thought experiment may be a powerful tool in showing which beliefs are shared and which are not shared, appearances notwithstanding. More generally, our analysis of two well-known successful thought experiments and our discussion of Mach led us to formulate the hypothesis that a thought experiment works under the following conditions:

I. At issue in the first condition is simply setting off the domain in which one is situated. A thought experiment deals with beliefs, not with things, nor with these mental representations of things that are supposed to be intuitions.[24] There is nothing to wonder at in this: any experiment whatsoever tests beliefs, the difference between a thought experiment and a physical experiment lying not in what is tested but in the way the test is performed.

II. A thought experiment deals with beliefs, not with things, which in our examples we called "assumptions" and which can perhaps more properly, insofar as there are not always explicit assumptions, be designated as the "context" of the thought experiment.[25] There is nothing to wonder at in this. The existence of shared assumptions or of a background context is a condition for argumentation in general.

III. These beliefs should be organised as it were in a hierarchical structure, and this hierarchical structure should be acknowledged by interlocutors, so that, in case of an inconsistency, they will agree to judge that it is better to abandon one belief rather than another.

---

[23] It is sufficient to refer readers to Pabst Battin, *Ethical Issues*.

[24] Kuhn, "Function for Thought Experiment," 242, 261, first pointed out that thought experiments test beliefs ("conceptual" or "mental apparatus" in his own terms)—which, as he was careful enough to point out, does not mean that they have nothing to do with the state of affairs in our world: "from thought experiments most people learn about their concepts and the world together" (253).

[25] Popper, *Logic*, 444, demands that idealisations are "concessions to the opponent or at least acceptable to the opponent"; Rescher, *What if?*, 9–14, insists from the outset on the need for a context.

It turns out that this is the main condition for a successful thought experiment.

3. *Unsuccessful Thought Experiments*

The examination of thought experiments that do not work will now allow us to argue that each of these conditions is individually necessary for a thought experiment to work. We will use simple examples to make clear that when we eliminate one of these conditions, we have a thought experiment that does not work. We will thus successively eliminate the condition that what a thought experiment tests are beliefs, the condition of agreement as to beliefs, and what is even more interesting in our minds, the condition of a hierarchy of beliefs. As above, we chose examples written at different periods and taken from various disciplines in order to point out that periods and disciplines are not pertinent parameters in our argument.

3.1. *What If We Eliminate the Condition (I) That What We Are Testing Are Beliefs?*

We will illustrate this case with the Cartesian thought experiment of receding bodies. Its domain is what Descartes calls "general physics," the part of physics that gives *a priori* demonstrations about matter and motion. In the article of the *Principia philosophiae* where the thought experience is carried out, Descartes wants to prove that the nature of matter, that is of bodies in general, is to be an extended substance. He must consequently eliminate all the qualities that we experience in bodies. The thought experiment concentrates on the difficult case of hardness and aims at refuting the atomist claim that hardness belongs to the nature of bodies.

> As for hardness, the only thing that our senses indicate to us, is that the parts of hard bodies resist the motion of our hands when they touch them; however, if whenever our hands moved in some direction, all bodies existing there receded with the same speed as our hands approached them, we would never feel hardness. But it is impossible to admit that bodies that recede in that way would for that reason lose the nature of bodies. By the same argument it can be shown that weight and colour and all other qualities of that sort that we sense in a material body can be taken away from it, leaving it intact.[26]

[26] *Principia philosophiae*, II 4, Descartes, *Œuvres*, VIII/1: 42, our translation.

In order to understand how Descartes deals with this problem, and what makes his thought experiment somewhat queer, let us first make explicit his system of beliefs:

(1) We are in the A-world.
(2) In the A-world, we touch bodies.
(3) When we touch bodies, resistance is felt.
(4) A body that offers resistance to motion is a hard body.
(5) Hardness belongs to the nature of bodies.
(6) We have knowledge about the nature of things in general (and bodies in particular).
(7) The nature of X is what X cannot be conceived without.
(8) The nature of bodies is independent from what we can feel about bodies.

In this consistent set, there are quite different kinds of beliefs. (4) and (7) are definitions, the first one derived from a phenomenological experience of a physical property, the last one specifying the meaning of a metaphysical term. As many metaphysical definitions, (7) has an unquestionable modal content, that makes this thought experiment undeniably modal. However, as we explained earlier, we are not so much interested in identifying types of thought experiments, which would imply relying on modal logic, than in showing how a thought experiment works. (2)–(4) are derived from our experience of bodies. (6)–(8) are metaphysical statements. To carry out the Cartesian thought experiment is to introduce the negation of (1) and (2): we emigrate to a W-world, in which, contrary to what happens in our A-world, we do not touch bodies because they recede from us. What happens in that case? Would we say that hardness belongs to the nature of bodies? It could be expected that, in a W-world, we would have other beliefs about bodies, for example, we can imagine that W-hardness would be, for W-people, the rate of speed with which bodies recede from their hands. If it were the case, the experience of W-bodies in the W-world would lead W-people to a different definition of hardness, so that (5) would be perhaps conserved, but with quite another meaning, tied to the different experimental contributions that would be expressed in the form of beliefs (W-3) and (W-4). However, Descartes claims that, in the W-world, we keep our beliefs (3) and (4), notwithstanding the fact that they are derived from our experience of A-bodies. Consequently, his test of consistency and his procedure of prioritisation of beliefs operate only on the subset (5)–(8). Now, for him, (6)–(8) are metaphysical statements that take precedence over (5): it is less costly to abandon (5) than to abandon (6)—which means being a sceptic as to the knowledge of the nature of bodies—or (7)—which means giving up a well-established definition of what is a nature—or (8), that can be seen as a consequence of (7).

The receding bodies thought experiment would most probably not have been accepted by Descartes' opponents, the 17th century atomists. They would have contested (4), because what they attribute to bodies was not the kind of phenomenological or sensible hardness Descartes talks about here, but an essential impenetrability defined as the property of a body to prevent any other body to occupy its place. However, it is another question that we would like to raise here—namely the question of why Descartes believes that in the W-world, beliefs (2)–(4) would not significantly change. It is obviously because his goal is to examine the nature of a body. Inasmuch as the nature of X is defined by (6) that without which X cannot be conceived, examining the nature of X means declaring the condition of possibility not of X, nor even of our experience of X, but of our conception of X. In the case of bodies, this means setting aside everything that is felt (for feeling is not conceiving), and keeping that without which neither bodies nor their properties can be conceived, that is extension.

If the Cartesian thought experiment had only the goal of defining the conditions under which bodies can be conceived, in other words, to use the terminology generally adopted in this article, making explicit the conditions under which we can have beliefs concerning bodies, we would stop at this point and say that we cannot conceive of bodies or of their properties, potentially sensory properties, independently of extension. However, a few lines earlier, at the beginning of Book II, Descartes warned us. The fact from which one should start is the existence of an extended substance that has all the properties that are clearly perceived to belong to an extended substance, this extended thing being what is usually called body or matter.[27] Hence, what Descartes means to speak of in Book II is not only the nature of bodies, but the bodies that exist; consequently, when he presents the receding bodies thought experiment, while the question is one of eliminating hardness and other sensory qualities, it is with reference to bodies that exist in our world. In other words, Descartes plays both sides of the fence: his thought experiment

[27] *Principia philosophiae*, II 1, Descartes, *Œuvres*, VIII/1: 41.

allegedly teaches us something both about the conditions under which a body can be conceived, and about the bodies of our world.

Indeed, this is a general problem in Descartes, not a problem specific to the receding bodies thought experiment. But we think that it is particularly striking in this case and that it is a nice illustration of the fact that we go off track if we drop the condition that thought experiments are exercises in reasoning about our beliefs and replace it with the idea that they are about the things themselves.

3.2. *What If We Eliminate the Condition (II) of Agreement on Beliefs?*

In general, thought experiments help us clarify our belief systems, revealing gaps or avoiding inconsistencies in them. Being exercises in reasoning, they have their strong points, but their weak points as well. In particular, it can happen that thought experiments fail to hit on a shared belief, be it a common experience or more generally a shared assumption. This might have been the case if Salviati had conversed not with Simplicio, but with an Aristotelian who would have refused the assumption that speed is mediative. This might have been the case as well if Thomson's opponents had been unable to agree with her on the language of rights. In these imaginary cases, Salviati's and Thomson's opponents might have begun to contrive counter-thought experiments and might have failed to find common ground. Examples of this kind of an out-of-control thought experiments-machine may be found in most controversies, but are especially striking in what Norton aptly called the thought experiment/anti-thought experiments pairs. As already noted by others, this is the case of the two Marys, brilliant scientists who, from birth, were confined to a laboratory, where they had only black and white experiences, but learnt everything about colours and colour-perception. Mary Jackson experiences something new when she leaves her laboratory and sees red for the first time; Mary Dennet however, when she leaves her lab, just exclaims: "Ah! Colour perception is just as I thought it would be!"[28] Exchanges of such thought experiments will quickly look like a fool's game, because they do not constitute *per se* a reason to choose one clear, complete and consistent system of beliefs rather than the other.

[28] Norton, "Why Thought Experiments," 45–49; Atkinson and Peijnenburg, "Thought Experiments Poor," 308–315.

3.3. *What If the Condition (III) of Hierarchy of Beliefs Is Eliminated?*

We will illustrate this case by an argument of counterfactual style that comes from Aristotle's *Politics*:

> ... if every tool could perform its own work when ordered, or by seeing what to do in advance, like the statues of Daedalus in the story, or the tripods of Hephaestus which the poet says 'enter self-moved the company divine'—if thus shuttles wove and quills played harps of themselves, master-craftsmen would have no need of assistants and masters no need of slaves.[29]

We will pass over quickly two preliminary difficulties. The passage quoted follows a formula that presents a translation issue; Aristotle has just spoken of the slave as an *organon pro organôn*. The formula can mean "a tool which serves for several tools" or else "an instrument that comes before the other instruments." According to the first interpretation, Aristotle wanted to oppose the specialisation of the instrument to the polymathism of the person using it; according to the second, he wanted to oppose the passivity of the instrument to the activity of the person using it. A second difficulty is that we cannot put on the same plane shuttles and harps. The first are instruments of *poiêsis*, the second instruments for the *praxis*: they thus refer to extremely different types of action. The interpretation that follows shows how we have chosen to resolve this conflict. Let us ask ourselves: "what are Aristotle's beliefs that the counterfactual supposition is supposed to undo?" Let us admit that the harp here is emblematic of instruments belonging to the order of the *praxis*. We would then have something like:

(1) An instrument is a thing able to accomplish the work that is proper to it.
(2) For an instrument, accomplishing the work that is proper to it, is to operate.
(3) Because it is unable to move itself spontaneously and even less able to foresee what is expected of it, an instrument is unable to operate by itself.
(4) In order for an instrument to operate, there must be an instrument of instruments.

---

[29] Aristotle, *Politics* 1253b17. We thank Frans De Haas for his sharp and constructive criticisms of our first presentation of this example.

(5) When instruments of instruments are human beings,[30] they are, depending on the case, slaves or workers.[31]
(6) Slaves and workers are in the service of masters and craftsmen respectively.

These are beliefs that Aristotle himself would use in a complex argument aimed at establishing that the master-slave relationship is natural. But let us consider them in and of themselves: they advance definitions [(1) and (2)]; they enlighten the relationship between craftsman and assistant or that between master and slave [(5) and (6)]; they tell us what are the negative [(3)] or positive [(4)] conditions that must be in effect for an instrument to be operational. Now it is clear that the thought experiment in this passage invites the reader to consider what would happen if the negative condition for the operability of instruments was modified. What would happen if the instruments became capable of moving on their own, or understanding what was expected of them? In other words, what would happen if there was a non-(3)? With the assumption that non-(3) is the case, (4) and (5) and (6) fall one after the other, like the legendary dominos: if an instrument can operate by itself, it is no longer indispensable that instruments of instruments make it operational, and it thus becomes futile to wonder what is the status of these instruments of instruments, whether in the field of *poiêsis* or in that of *praxis*. In short, the introduction of non-(3) does not allow us to identify a weak link that we could then eliminate: it jeopardises Aristotle's entire set of beliefs,[32] which is why he prefers not to reorder them and ends up refusing to admit non-(3) as absurd. The reason is that the hierarchy among these beliefs is not clearly determined: if one goes down, the other goes down with it: united they stand, divided they fall.

Conclusion

As a conclusion, we will make four remarks: first, on the incomplete character of our argumentation; second, on the domain of validity of our description of the mechanisms of thought experiments, with respect to

---

30 These could also be animals, but we will set aside this point.

31 Depending on whether one is dealing with instruments of *poiêsis* (such as shuttles) or of *praxis* (such as zithers). On zithers as instruments of *praxis*, see *Magna moralia*, 1997a9–15.

32 But not, however, (1) and (2), which give prior definitions of the instrument and the operation that remain valid, whether the instruments can move themselves or not.

different historical periods and different fields of knowledge; third, on our position in the contemporary debate on thought experiments; and fourth, on the fact that our position does not imply any kind of general relativism.

i) In the course of our argumentation, we argued through counter-examples that the three conditions (I), (II) and (III) were individually necessary, not that they were conjointly sufficient. To proceed by induction to demonstrate this would be tedious, not to say impossible, and our paper does not meet the conditions for a rigorous proof—in particular because, even if we admitted the broad characterisation of thought experiments advanced in the introduction of this book, we do not have a clear-cut definition of the notion. It is thus only as a conjecture that we will advance the thesis that the conditions (I)–(III) are conjointly sufficient, a thesis we will hold to be true until a counter-example can be opposed to it.
ii) The conditions (I)–(III) are conditions on beliefs that are not proper to a historical period or a field of knowledge.[33] But we can understand retrospectively why they are easily met in the early mechanics of the 17th century analysed by Mach: operations on bodies, some more elementary than some others, had led to the constitution of a reservoir of shared and hierarchised beliefs. We also understand that thought experiments are usually effective in a field like physics, where there are principles or laws upon which every physicist agrees, and which trump all other propositions. We finally understand what happens in these out-of-control thought experiments-machines that sometimes emerge in physical controversies. But we do not see why there would be a sharp essential difference between thought experiments in the hard sciences and in the other fields of knowledge. There is no *a priori* reason for our conditions not to be met elsewhere: this is merely a *de facto* question. They were met in the case of Thomson's thought experiment dealing with beliefs related to propositions concerning rights: this does not mean that it is the case in the whole domain of applied ethics. Thought experiments appear much more difficult in the present state of economics than in physics, but this state of affairs does not rule out the

[33] This was already a background assumption of Atkinson and Peijnenburg, "Thought Experiments Poor." On the contrary, Wilkes, *Real People*, and Rescher, *What if?*, argue that there is an essential difference between philosophical and scientific thought experiments.

existence of thought experiments in a particular economical school or, in the future, in economics in general.[34]

iii) With respect to the existing literature, the description we just proposed of thought experiments inclines us, without committing us, to a deflationary position in at least two respects. As we noted to begin with, our description of the mechanism of thought experiments does not commit us to any position concerning their nature. In particular, we do not need the strong deflationist thesis that thought experiments are only arguments; it is sufficient for us to admit that they are arguments. However, we have to confess that we have more sympathy for the deflationist thesis of Norton, according to which thought experiments are only arguments, albeit with some qualifications, than for the imaginarist thesis of Gendler that a thought experiment contains something more than an argument. Even so, our aim was not to defend the deflationist thesis as such, and it is only as a conjecture that we will advance the thesis that thought experiments are arguments, a thesis we will hold to be true until counter-examples can be opposed to it.

There are mainly two reasons for which we are not convinced by Gendler's thesis. First, we think that she was induced in the first place to defend such a thesis because of her excessive expectations with regard to arguments and, more generally, because of her most unrealistic vision of the sciences. In her paper on Galileo, she writes, for example, that "to draw a conclusion on the basis of a *non-thought-experimental argument* is to be led by a process of inductive or deductive reasoning from a set of explicit premises."[35] Her argumentative strategy is therefore the following: inasmuch as the strapped-bodies thought experiment is not totally convincing without "broad, defeasible, tacit assumptions," that are not explicit in Galileo's dialogue, something should stand in for them.[36] But, contrary to what she implies, no real experience in physics would be convincing, if background knowledge were not mobilised. This does not make physical experimentation mysterious: a certain num-

[34] Cartwright, "Vanity of Rigor," assimilates theoretical models in economics and thought experiments in physics, because both "seem like toys. They have no texture to them, no tones of colour, no gradations. They are simple situations, with caricature objects with few options for their behaviour" (217–218). Her argument, quite different from ours, relies on a confusion between idealisation and thought experimentation.

[35] Gendler, "Galileo," 399.

[36] Ibid., 406–408. For a detailed criticism of Gendler's position, see Zeimbekis' essay in this volume.

ber of these implicit premises are made explicit when they are contested, for example during controversies. Second, Gendler does not succeed in clearly specifying what would be this supplement that would help understand the outcome and justification of a thought experiment. She speaks of "quasi-sensory intuitions" or "mental images" and invokes acts of cognition that do not result from explicit calculating or reasoning, like imagining how many elephants can fit in a room, or whether this carpet would clash with bright green painted walls.[37] Such acts of cognition indeed exist, as the literature she referred to amply shows,[38] but it is not at all certain that they are actually involved in thought experiments, which do not appear on the mental scene of an individual, but in highly sophisticated controversial contexts, where a choice between competing hypotheses or even competing theories is to be made.[39]

We consequently, if only because of a general principle of economy of thought, feel closer to the deflationary thesis defended by Norton. Note however that Norton, inasmuch as he was answering Brown,[40] explicitly links together his deflationary thesis and his empiricism: if thought experiments were not merely arguments, empiricism would be endangered by "a mysterious route to knowledge of the physical world," "epistemic miracles" or "epistemic magic."[41] While we are ourselves more empiricists,—although we would have to explain in which sense—, we think it is important to note that the reciprocal proposition (empiricism implies the thesis that thought experiments are merely arguments) does not seem to us to be assured. The analysis inspired by Rescher that we have proposed to specify the mechanism of thought experiments seems to us to be compatible with all sorts of positions in terms of theory of knowledge, not only with empiricist positions. At the risk of repeating ourselves, materials providing input in thought experiments come out of beliefs that can have all sorts of origins.

iv) Speaking of "shared beliefs" can evoke certain relativistic theses, if only because this expression is sometimes used by the tenants of the

---

37 Gendler, "Thought Experiments Rethought," 1156–1159.

38 Ibid., 1159–1161.

39 Norton, "Why Thought Experiments," 60–61, criticises on similar grounds the mental modelling views of thought experiments.

40 Brown, "Thought Experiments"; idem, "Why Thought Experiments Transcend," 30–31.

41 Norton, "Thought Experiments," 129; idem, "Are Thought Experiments," 334–337; idem, "Why Thought Experiments," 44–45.

> thesis that a shared belief is *per se* a good belief. Our position with respect to relativism is simple: we are contextualists as to thought experiences, but we do not feel that this local contextualism has any reason to lead to relativism, not to speak of general relativism.

We can perhaps make ourselves understood by a little thought experiment of our own—which, at this point of our paper, is rather an illustration. Let us suppose that a group of individuals share the following beliefs: there is a very bad being; this very bad being has three and only three hairs; of these three hairs, one is blue, the other yellow, the last one red. Suppose moreover that these three beliefs are assumptions so fundamental for this group of individuals that they have developed, without ever doubting them, fairly sophisticated theories based on them. If these individuals enjoy arguing as we do, they have a set of beliefs that meet the three conditions we have defined: they will thus be able to put in place thought experiments that work, thanks to which one of them can convince others of some particular property of this very bad being. But now one of the individuals of this group, not only talkative but adventurous, leaves her companions to meet with a very different group: will she not perceive that the thought experiments she has laboriously put in place no longer work in this new group? This illustrates that a thought experiment does not work in and of itself, but in a certain context, but this does not call on a relativist thesis as such, but only on a minimal contextualism. Relativism does not consist in saying that there are shared beliefs in certain contexts, beliefs that are not shared in others; it is the thesis that shared beliefs, because they are shared, are good beliefs, and in particular that there is no epistemic criterion for choosing among opposing and competing beliefs. In other words, for a relativist, the best belief is the most widely shared belief, the nature of this "best" being consequently variable. In our paper, we take no position as to this relativist thesis.

However, even if we were relativists with respect to thought experiments, we do not see why this local relativism would necessarily imply a general relativism, extended to the physical sciences. What keeps us from being relativists in physics is not the conviction that there would be, if only once, a pure observation, a true law of nature, a concept corresponding to a natural kind: we know how problematic each of these affirmations is. It is, to say it naively, the fact that there are physical theories which, taken as a whole, work, insofar as, and without us needing to determine here why, they allow for a certain number of predictions, which is today generally not the case, for example, in history, in economics, in philosophy or in sociology. This is the reason why certain thought experiments have a predictive character in physics; their predictability coming not from the thought experiments as such, but from the fact that they occur within the context of predictive theories.[42] Thus, even if we were not only contextualists, but were relativists with respect to thought experiments, our relativism would *prima facie* not be extended to all the physical sciences.

[42] For a similar move, see Atkinson and Peijnenburg, "Thought Experiments Poor." They note, however, that unsuccessful thought experiments are less disastrous in science than in philosophy because, in science, we can turn to theories (and there are good reasons to prefer one theory over the other) or to real experiments (and there are crucial experiments in the sciences).

THOUGHT EXPERIMENTS AND MENTAL SIMULATIONS

John Zeimbekis

1. *Simulations, Thoughts, and Natural Processes*

There are two ways in which thought experiments could be mental simulations, depending on what kind of process is being simulated. Some thought experiments could be mental processes simulating *non-mental* processes; others could be mental processes simulating *mental* target processes. The second category would contain thought experiments which seek to acquaint us imaginatively with moral dilemmas, whether by engaging our reasoning or by replicating the epistemic conditions under which moral sentiments may be aroused. Such thought experiments could be called "mental-mental" simulations (mental source-processes, mental target-processes). Physical thought experiments, on the other hand, would be in the first category: they are conducted by means of processes (thoughts) which do not occur in their targets (external physical events), and could be termed "mental-physical" simulations.

A standard example of simulation is testing a model airplane in a wind tunnel (a *physical*-physical simulation). Although the variables which the simulation is about do not appear in the simulation, as is also the case in thought experiments, we nevertheless, unlike in thought experiments, delegate the task of determining the outcome of the natural target process to natural processes and not to thoughts. Accordingly, when we have to simulate mental target processes, simulations are conducted by mental processes. This is consistent with a central thesis of the simulation theory of mind: there has to be some "relevant similarity" between source and target processes for a simulation to be possible in the first place.[1] Since mental-mental simulations experiment *with* thoughts *about* thoughts, they are not substitutes for experimentation but genuine experiments—a status not normally granted thought experiments. But now, mental-*physical* simulations, which is what physical thought experiments would

[1] Davies, "Mental Simulation"; Stone and Davies, "Progress Report"; Stich and Nichols, "Folk Psychology."

be if they were simulations, would appear to be an oddity for simulation theory: thought experiments are about physical processes, but they are thought-conducted, so how can they be *simulations*?

Nevertheless, there have been several attempts[2] to explain both the mental execution of thought experiments and the principles underpinning their epistemological validity by using some concept or other of simulation.[3] These authors exploit a hunch already present in Ernst Mach, who claimed that during physical thought experimentation our thoughts "mimic" natural processes (see section 3). The concept of simulation, apart from requiring similarities between source and target processes, also implies *dissimilarities*—the dissimilarities between the variables appearing in the simulation and those whose behaviour the simulation is intended to predict. It is by exploiting this aspect of simulation that theories of thought experimentation attempt to explain how thought can inform us of the workings of something as dissimilar from it as physical processes. At the same time, simulation theory itself offers some encouragement to such attempts, for it, too, sometimes seeks to bridge the gap between dissimilar processes or phenomena by defining special forms of epistemological access. For example, some simulationists describe mental visualisation as a substitute for seeing, which is potentially useful for bridging the thought/observation gap in physical thought experiments. Others describe imagining—construed not as a form of mental representation, but as a special mental attitude adopted *towards* mental representations—as a substitute for belief.[4] This is useful for analysing thought experiments which seek to reconstruct the cognitive underpinnings of motivational and emotional states.

Apart from such explicit appeals to simulation, theories which claim that thought experiments are not reducible to bare arguments also potentially leave room for simulation to play some role, since something else has to stand in for arguments according to such theories. Besides, in certain respects, thought experiments and mental simulations are quite similar. Both are mental activities in which (a) we mentally represent a hypothetical or a counterfactual set of circumstances, and (b) we mentally process that *representation* in order to reach a prediction about how the set

---

[2] Gendler, "Thought Experiments Rethought"; Gooding, "What is Experimental"; Nersessian, "Cognitive Basis" and "Model-based."

[3] Currie, "Visual Imagery"; Nersessian, "Cognitive Basis"; see also Walton, *Mimesis*, Ch. 8.

[4] Nichols, "Imagining and Believing"; Walton, *Mimesis*; Currie and Ravenscroft, *Recreative Minds*.

of circumstances it represents would behave. In each case it is required that, without conditions actually obtaining as they would in an experiment, we can nevertheless, by using only our mental resources, make a prediction about those conditions.

Of course, the mental process by which we reach a prediction about the outcome of a system's physical state may be *neither* an argument *nor* a simulation. It may be an intuition of some as yet unspecified kind, or some (yet to be specified) use of prior implicit knowledge. My purpose here is precisely to clarify these options and choose between them by distinguishing the respective roles of: (a) process-driven simulation, (b) mental modelling, (c) visualisation, (d) implicit knowledge, (e) induction, and (f) inference. Unless this is done, the term "simulation" (like "intuition") will be used in as broad a sense as possible (covering (a) to (e) in the preceding list), and will merely serve to re-baptise, without dispelling, the mystery of how thought experiments can gain access to the workings of the thought-transcendent world.

2. *The Relevant Similarity Constraint*

The relevant similarity constraint on source and target processes applies only to *process-driven* simulations, or to simulations to the extent that they are process-driven.[5] This leaves room for a concept of *theory-driven* mental simulation, which does not require relevant similarity because it merely involves application of prior theoretical knowledge to what we are imagining. This would occur if, for instance, in Stevinus' thought experiment, a mental representation of the chain revolving perpetually did not conflict with any implicit knowledge (such as domain-specific know-how manifested in sensorimotor responses), and we had instead to reach the prediction that the chain does not revolve perpetually by applying an acquired theory that there is no perpetual motion. The view that thought experiments are *theory-driven* simulations would be compatible with Norton's view that thought experiments are "merely picturesque arguments" and that the knowledge of the natural world afforded by thought experiments comes "from premises introduced explicitly or tacitly into the thought experiment," which is then "transformed, usually tacitly, through deductive or inductive argumentation to give the final

[5] Davies, "Mental Simulation"; Stone, "Progress Report"; Stich and Nichols, "Folk Psychology".

result."[6] Similar reservations about the possibility of process-driven simulation (and support for theory-drivenness) are expressed by Dennett even for *mental* targets during the activity of mindreading:

> If I make believe I am a suspension bridge and wonder what I will do when the wind blows, what 'comes to me' in my make-believe state depends on how sophisticated my knowledge is of the physics and engineering of suspension bridges. Why should my making believe I have your beliefs be any different?[7]

If there was no relevant similarity between target and source processes in the putative simulations of thought-processes, thought-simulations would have to proceed by applying generalisations about situations to predict human behaviour. But against Dennett, simulationists argue, plausibly enough, that there *is* relevant similarity because some mental processes "operate in just the same way when we imagine being in a particular situation as they would if we were really in that situation."[8] Indeed, Dennett's own theory does not suppose that mindreading proceeds by treating individual cases of practical reasoning by applying acquired theoretical generalisations—something which would imply a strange notion of practical rationality. It instead holds that mindreading proceeds by applying normative rational constraints. This has given rise to Jane Heal's simulationist interpretation of Dennett's normative intentional stance.[9] If the target mind uses counterfactual reasoning and applies some normative concept of belief to reach behaviour from beliefs and desires, my own mind will use the same processes to reach imagined behaviours from suppositions—because, and to the extent that, we can assume that my own counterfactual reasoning, concept of belief, semantic constraints, and so forth, are relevantly similar to the target subject's. I use my own mind to see what would happen in another mind.

This idealised description of mental simulation in mindreading shows what a clear case of relevant similarity between mental processes and targets *would* look like, and on what basis mental simulations *could* succeed. It shows that a mental activity is substantially a simulation only if it is process-driven, that is, *only if it simulates the processes by which the target goes from an initial state to a subsequent state.*

---

[6] Norton, "On Thought Experiments," 1140.

[7] Dennett, *Intentional Stance*, 102.

[8] Davies and Stone, "Simulation Theory."

[9] Heal, "Understanding Other Minds"; this is the sense in which I apply the concept of simulation in sections 4 and 5 below.

This condition is not met by everything we call a simulation in other senses of the term. A mental representation may be said to simulate what it represents in the weaker sense, that it resembles or replicates spatial and temporal relations between the parts of the target, while at the same time, the *way* we reach predictions about the target's transition from one state to another is not by simulating the referent's processes. There is certainly no *prima facie* similarity between thought processes and processes which determine the physical states of external objects. This is implicit in existing formulations of the epistemological problem of thought experiments: "Thought experiments are supposed to give us knowledge of the natural world. From where does this knowledge come?";[10] "How is it possible to learn apparently new things about nature without new empirical data?"[11] For, where mental simulation *does* work, in the mental simulation of thought processes, two kinds of reply to these questions spring immediately to mind. (1) We learn new things about nature by using our thoughts because we experiment with our brains ("thoughts") to learn about other brains ("nature"). (2) We learn something about mental phenomena by experimenting with mental phenomena. In either case, the reply is available because it is in principle possible to claim some form of similarity between target and source processes. The problem is that relevant similarity does not seem to obtain other than in thought simulations of thought processes: how can experimentation processes which are not designed for physical-level predictions and explanations, but for mental predictions and explanations, make physical-level predictions and still be experiments?

3. *Physical Thought Experiments*

Attempts to respond to this problem in theories of thought experiments can be described as theses about *special epistemological access*. By this, I refer to a growing set of claims concerning the role in thought experimentation of various components of simulation theory, especially visualisation and perceptual-motor activity in the manipulation of mental images, as well as to forms of non-propositional knowing, knowing-how, domain-specificity, or modularity.

---

[10] Norton, "On Thought Experiments," 1140.

[11] Brown, "Thought Experiments."

Such appeals can already be found in Mach's concept of a form of unarticulated "instinctive knowledge" derived from the observation of natural processes. Two traits of this instinctive knowledge of physical processes described by Mach are especially relevant to our purposes: (i) it "exists in absolute independence of our participation"; and (ii) it is "imprinted" "in our percepts and ideas, which, then, in their turn, mimic the processes of nature."[12] The first characteristic implies that the implicit knowledge enables us to have certain expectations about the outcomes of physical processes as a result of processes which occur spontaneously, as autonomous events independent of our agency. This suggests that the ability which leads us to expect certain outcomes from imagined physical processes is sub-doxastic and non-inferential. If this is so, then the predictions drawn from thought experiments cannot result from the application of laws or other generalisations, and therefore are not theory-driven but process-driven. What remains to be seen, however, is whether this process is a simulation in the first place.

Mach's claim that perceptions *mimic* the processes of nature gives this impression. The claim can at best be cashed out as meaning that certain mental representations trigger affordance-based, domain-specific, perhaps modular, reactions to physical processes. But the "mimicry" or replication in question will not be a replication of target physical process by cognitive or brain processes. When we perform Stevinus' thought experiment, not only do we not have a scaled-down prism-and-chain system in our heads, but we *cannot* have anything relevantly similar in respect of causal properties to the prism-chain system: for if the model were relevantly similar, then it would be nomologically impossible for the chain-analogue to slide around the prism-analogue in the way we can *represent* it as doing.

So Mach's concept of mimicry cannot imply a replication of the structural properties of the target system. What, then, is intended by the use of the concept? Surely, the fact that just as before a real chain on a prism we would be surprised to see it slide, we will be *similarly* be surprised before a spatial mental representation of a chain sliding around a prism. But all that this means is that our affordance-based dispositions towards physical objects can be triggered by imagining situations as well as by seeing situations. The replication or similarity relation obtains only between our *responses* to *F*s and our *responses* to mental representations of *F*s, not

[12] Ernst Mach, *Science of Mechanics*, quoted by Sorensen, *Thought Experiments*, 54, 51.

between mental processes and *F*s. These responses are enabled jointly by the mental representation, taken as an initiating cause, and by domain-specific know-how construed as a disposition. The initiating cause (the spatial mental representation) bears a perceptual similarity to the target, but neither the inner cause (the implicit knowledge of the target process), nor the manifestation of the disposition (the expectation that Stevinus' chain will not revolve) bears any similarity to the target process. Since it is the inner cause which determines the imagined outcomes, similarity need not play any role in determining which imagined outcome we will accept. The similarity between visualising and perceiving the target is due to preservation of spatial and topological features. But even iconic representations, as in fact Plato pointed out, bear only an appearance-based similarity to their depicta, not a similarity in respect of observer-independent, world-to-world causal properties. Mach's example actually brings clearly into focus the fact that there cannot be a replication of target-properties in the mind or the brain, because the thought experiment affords us knowledge by mentally visualising a physically impossible, but logically possible, event—something which can be done only by *avoiding* relevant similarity, that is, representationally.

One author who applies Mach's ideas by using current theories of psychology and philosophy of mind is Tamar Szabo Gendler. It is worth comparing the question that Gendler formulates in this connection with the question formulated here—namely, how experimentation processes designed for mental-level predictions can be expected to make physical-level predictions. Gendler asks "how contemplation of an imaginary scenario can lead to new knowledge about contingent features of the natural world," and, significantly, sees this question as "a special case of a more general one, namely how any *nonperceptual* capacity can lead to new knowledge about (nonstipulated) contingent features of reality."[13] This sets the course for her reply, which assimilates thought experimentation, construed as "reasoning about an imaginary scenario," to the perceptual observation involved in ordinary experiments. The reply draws on the idea that affordance-based sense, proprioception, and manipulation of mental images generate new information in a non-argumentative way to acquire "knowledge of contingent features of the natural world." Gendler illustrates such reasoning with examples such as trying to figure out how many elephants can fit into a remembered

[13] Gendler, "Thought Experiments Rethought," 1152; italics added.

room, and whether it is possible to cycle around an imagined room with obstacles without tilting. The comparisons assimilate the imaginings of physical thought experiments to the observation of physical experiments by using as a middle term for the comparison mental processes which have one foot in the mind and one in the physical world, such as the non-conceptual and proto-conceptual contents typically discussed in philosophical theories of perception.

To bring out this point, consider Dretske's concept of simple or non-epistemic seeing: I may not have formed any beliefs about how many children were playing in the courtyard when I saw them, but I may be able to retrieve the information subsequently if I'm asked, by remembering the scene and picking out a plausible number of figures in it.[14] Dretske's theory exploits the fact that the information conveyed causally, "naturally," during perception is dense and analogue before being conceptually and propositionally encoded in the formation of beliefs. Gendler's theory of thought experiments exploits similar traits of mental imagery construed as a store of information about the world, only that Gendler additionally concentrates on the active, sensorimotor uses of such non-conceptual information: I possess information, in some quasi-natural form, about the natural world; retrieve that quasi-perceptual information; interact with it in a sensorimotor and affordance-based manner; and draw conclusions about the physical world on the basis of those interactions. So Gendler's answer to our question about how we can experiment with thoughts about physical processes is that, at the relevant moments, "thought experiments" are not *thought* experiments but are *quasi-physical* experiments.

To illustrate the possibility of acquiring such non-inferential knowledge, Gendler presents an experiment on mental imagery by Daniel Reisberg.[15] In Reisberg's experiment, subjects are (a) shown a form, (b) asked to memorise it, (c) to imagine it rotated, and (d) to draw a picture of the rotated form. When presented with a familiar geographical form rotated by 90 degrees, no subject succeeded in recognising the form by rotating the image mentally (step c), but many were able to recognise it in their own drawing (step d). Gendler points out that these subjects have now acquired a "new justified true belief (that the rotated image resembles Texas)" without "inductive or deductive reasoning from known

[14] Dretske, *Knowledge*, 135–153.
[15] Reisberg, "Non-ambiguity."

premises."[16] The following comparison is then made between the mental procedures involved in the Reisberg experiment and those involved in Stevinus' thought experiment:

> What's important for my purposes is the extent to which this case [Stevinus'] resembles those described above [Reisberg's]. Contemplation of an imaginary scenario (the cut string laid atop the prism) evokes certain quasi-sensory intuitions, and on the basis of these intuitions, we form a new belief about contingent features of the natural world (that the weight of four balls offsets the weight of three balls). This belief is produced not inferentially, but quasi-observationally: the presence of the mental image plays a crucial cognitive role in its formation.[17]

It is true that in both cases, Stevinus' and Reisberg's, belief-formation requires the presence and manipulation of a mental image. But as we saw in the analysis of Mach's concept of mimicry, this could be a trivial truth about mental representation, not a substantial claim about what determines the outcomes of thought experiments. In this latter respect, the Stevinus and Reisberg cases do *not* resemble each other. In Reisberg's experiments, the necessity of mental rotation for recognition is shown because of similarity between source-processes and target-processes: the experiment is *about* a mental process, visual recognition, and it is *conducted by* a mental process, rotation of mental images. Since the pictures preserve the features which are required for object-recognition procedures to operate, the relevant similarity constraint is met: it could be said that the picture-perceptions are simulations of object-perceptions. This contrasts starkly with Stevinus' thought experiment, which, while it is conducted mentally, is about a physical process whose relevant features it cannot replicate. Note what we *can* say about the Stevinus case: that the mental visualisation of the chain on the prism resembles the perception of a chain on a prism in many relevant respects. But the respects in question are perceptual, and the thought experiment is not about the *perceptions* of chains on prisms, it is about the causal properties *internal* to systems comprising chains and prisms. The behaviour of those internal properties is *predicted* neither by process-driven simulation nor by contemplation and manipulation of the mental image, but by inductive application of prior implicit knowledge *to* the mental image as if it were not an image but an object.

[16] Gendler, "Thought Experiments Rethought," 1161.
[17] Ibid.

How does this square with Gendler's conception of non-inferential knowledge? In the passage cited, the key claim lies in the concept of *quasi-observationally derived belief*. The concept is not only contrasted to inferentially derived belief, but has also to designate what *replaces* inference; and since there is no actual observation, it is the expression "quasi" in "quasi-observational" which bears the weight of the explanation. Consider first the following example of a "quasi-*F*" which appears to be efficacious in the framework of a simulation. During our psychological responses to fictions, we sometimes experience emotions which are qualitatively similar to emotions that we experience in real contexts, but which lack the cognitive causes of real emotions, namely, beliefs. Quasi-emotions make fictions useful for simulating real emotions; and since emotions are among the springs of action, this in turn should make fictions useful for predicting actions in hypothetical contexts. Thus, in the expression "quasi emotion," "quasi" means two things. First, it means "not a real *F* but similar to an *F* in its phenomenal quality." Second, since the phenomenal quality of an emotion is a causally relevant property of an emotion, this implies that a process which can instantiate that property will bear a relevant similarity to the process which causes real actions. So by virtue of their similarity to real emotions, quasi-emotions confer on the experience of fictions the *capacity to simulate the processes by which action is determined*, and by virtue of this, *the capacity to predict action*.

Now, when we state that in thought experiments beliefs about the physical workings of the natural world are *quasi-observationally* derived, do we mean that there is some mental process which, once put into action, can simulate the outcomes of physical processes by virtue of a relevant similarity to physical processes? No, because in the expression "quasi-observational," "quasi" applies to the similarity between visualisation and perception, not to any similarity between on one hand physical processes, and on the other, the process by which *from* certain initial visualisations (inputs) we reach *further* visualisations (outputs). This contrasts with those cases, mentioned by Gendler in support of her argument, in which individuals overcome their fear of flying by imagining flying safely.[18] Gendler's point is that in such cases we form a new belief that flying is safe, not on the basis of deduction or induction, but just on the basis of imagining. These cases closely resemble the simulations just described involving quasi-emotions, and they succeed because both the

[18] Gendler, "Thought Experiments Rethought," 1160.

source and target processes are mental. But for that reason, they cannot solve the problem of physical thought experiments, which would have to be mental-physical simulations.

Gendler's examples bring into focus a further difference between thought experiments and simulations which should be a source of worry for the simulation hypothesis on thought experiments. The example in which we imaginatively fit elephants into a living room involves remembering a *particular* room and the spatial properties of members of a *particular* natural kind. This reminds us that simulation is well suited for rehearsing the behaviour of targets with *determinate spatial properties*, and therefore for dealing with contingencies to which physical laws are blind. This conflicts with physical thought experimentation, in which we tend instead to abstract from contingencies. The usefulness of fine-grained non-conceptual contents, proto-propositional mental contents, short-term memorisation of the contents of simple perceptions for later retrieval, short-lived indexical concepts, and egocentric spatial representation, is context-relative. If we have an ability to mentally rehearse such events as getting the piano out of the window, it remains possible that that ability *as such* will not be suitable for extracting general conclusions about the physical world. The difference is worrying because it may mean that while we have an ability to reason spatially, and even to combine conceptual thought with perceptual or imagined contents, this ability may be exhausted by the production of *mental models* in physical thought experiments, without implying the presence of any form of process-driven simulation. An insufficient conceptual and cognitive analysis of thought experiments may lead us to confuse these processes of mental modelling with process-driven simulations. For example, "simulation" may just mean that we inductively project prior knowledge, whether implicit or explicit, to perceptual-style imagery *as we would* to objects of perception.

This is borne out by work carried out on mental modelling. Mental modelling is a widespread activity found not only in thought experiments but in the navigation and manipulation of physical objects, and in the formation, possession and revision of concepts. It seems to be the natural inheritor of Mach's project, but provides a clearer position on what determines imagined outcomes during the mental manipulation of imagined objects. Instead of appealing to process-driven simulations, it appeals ultimately to constraints which are embedded, usually implicitly, in *concepts*. Concepts are described by Nersessian as sets of constraints on generating occurrent mental models; so on this theory, it is possible that what constrains the relation between imagined inputs and imagined outputs in thought experiments is concept possession and mastery. The kinds of mental models used in physical thought experiments are defined as mental representations that preserve the salient spatio-temporal and causal relations of target objects.[19] Thus, suppose that I attempt to predict the outcome of bending a rod. The rod-representation will have to be of something isomorphic to a rod, not to a spring or to a stone, in order for the knowledge implicit in the constraints on my concept for a rod (rather than my concept for a spring or a stone) to be triggered and applied. "In order for" here introduces a causal relation between, on one hand, the structure of the mental representation, and on the other hand, the application of prior inductively formed knowledge. The operative processes are the isomorphic nature of the representation, and the inductively formed set of constraints on concepts. Thus, there is *no simulation of physical processes* occurring in the rod, but only a simulation in two other senses. (1) There is simulation *of a perception of* a physical state. This triggers (2), existing affordance-based knowledge about the outcome of the state, a process involving no simulation. (3) There is inductive application of that knowledge *as if* we were applying it *in vivo*. For imagined outputs to be determined by a process-driven simulation, we would have to reach them by submitting the input-representation to a causal process that bears metaphysical similarity to the causal properties of the rod. Not only *do* we not do this, it seems that we *cannot* do it—we cannot submit a mental representation of the rod to a process of flexion. This is, of course, evident; what is not evident to start with, and what I hope this makes clear, is that what decides the outcomes of physical thought experiments cannot be process-driven simulations.

4. *Mental Targets, Rationality, and Emotions*

The situation is altogether different where mental-*mental* simulations are concerned, because these simulations comply with the relevant similarity constraint. Moreover, the power of simulation to individuate cases by replicating contingent local features (which are irrelevant for formulating physical generalisations) suggests that it is a good method for individuating action types and rehearsing practical reasoning. Practical reasoning typically does not draw conclusions from generalisations (in the form of

---

19 Nersessian, "Cognitive Basis," 141; Johnson-Laird, "Mental Models."

prima facie premises) and has to be sensitive to local contingencies. Our mental-mental simulation abilities are thought to have evolved for the purpose of mindreading, which is the understanding of the epistemic states of individuals in given contexts. As such, mental-mental simulation also looks useful for rehearsing reasons and justifications in moral thought experiments, since these typically involve imaginatively placing subjects in concrete situations.

My main purpose in this section is to acquire the means to address the problem of *moral* thought experiments in the following section. Moral thought experiments are thought to acquaint us imaginatively with the epistemic situations of agents in moral dilemmas. So in this section, I will discuss the requirements for moral thought experiments so construed, by asking whether relevant similarity in fact obtains, and reviewing some problematic areas of mental-mental simulation that could be relevant to the moral cases. I will draw largely positive conclusions about the possibility and usefulness of mental-mental simulations. In the following section, however, I will argue against the usefulness of construing moral thought experiments as mental-mental simulations.

A key question about the processes required for mental simulations is whether, and how, we can simulate the causes of irrational behaviour, such as weakness of the will, using rational processes. Suppose that we adopt a simulation theory based on application of rational procedures and norms, such as that described in section 2. As it stands at least, such a simulation will lead to false predictions in any target area where non-rational processes play a causal role. When such breakdowns of rationality are failures in the physical mechanisms *implementing* rational procedures, they cannot be predicted by using *mental* simulation. Suppose that I try to predict whether a smoker will resist buying his next packet of cigarettes. To make a valid probabilistic prediction, we should take into account processes which override rational procedures altogether, such as chemical reward pathways created by habit. So we cannot reach a valid prediction simply by placing ourselves in an imaginary situation *as agents*, but have also to apply a theoretical generalisation. This problem cannot be avoided by construing the brain mechanism in terms of its conscious, phenomenal manifestation, namely, an urgent desire that we smoke. For *thus* construed, as a desire, the cause enters the realm of reasons and (under *ceteris paribus* conditions) we cannot but rationally reach the all-things-considered conclusion that the agent *prefers* to abstain. When norms of rationality are used to simulate, they tell us what an agent *ought* to do, not what he *will* do. To know what the agent *will* do, either the rationality-based simulation has to be completed with prior theoretical knowledge, or else some way needs to be found of simulating the causal role of states other than beliefs and desires.

A particularly important question in this respect is whether, and how, the causal role of *emotions* can be predicted by rationality-based simulations, because according to some authors,[20] emotions can contribute to causing actions independently of reasons by focusing attention on aspects of the agent's situation that reasons fail to capture. For present purposes, two possible claims have to be distinguished concerning the emotions. One is that emotions have *rational underpinnings*; the other is that emotions *comply with rational constraints* when they contribute to determining actions. A striking version of the first claim can be found in the theory that emotions are perceptions of values.[21] According to this theory, emotions stand to values roughly as sensations stand to objective properties: they are the form in which the practical values and affordances of objects and situations are signalled to us in experience. For example, the emotion of fear is a perception of danger; the concept *danger* is a thick evaluative concept with a negative polarity, and, *ceteris paribus*, it implies disvalue. On such a theory, emotions are rational in the sense that they exist in the first place for reasons which comply with our interests (assuming an initial set of priorities, or "system-objectives" in Dennett's terminology). If such a theory could be generalised, it may show that evaluations are already built into the *descriptive* contents under which we represent imaginary situations, and if this is so, the imaginings should cause at least a *representation* of the appropriate emotions. Now, even this strong form of cognitivism about the rational underpinnings of emotions does not mean that a *given* emotion has to comply with rational constraints when it contributes to determining an action. For example, sentimental or sexual jealousy may have rational underpinnings under the set of priorities, which initially set up our capacity to experience jealousy, so that jealousy may be said to exist in the first place to alert us to the presence of a certain kind of danger. All of this constitutes in an externalist sense a rational underpinning of jealousy, but it still does not mean that a particular episode of jealousy cannot be a cause of irrational action. The phenomenal character of jealousy, as distinct from its cognitive content which can enter propositional thought, can persist and cause a subject to act solely on its account, in a manner not too different from

20 Tappolet, "Akratic Action"; de Sousa, *Rationality*.
21 Mulligan, "Appropriate Emotions."

that in which the content of the smoker's desire can enter rational calculations, but not its motivational force.

However, the simulationist can appeal here to a body of evidence which suggests that the emotions can in fact be simulated, and to a concept of *imaginative acquaintance*: knowledge we acquire about situations which are described in sufficient detail from a subjective point of view, as generally occurs in literary or artistic fiction. For the concept of imaginative acquaintance to work, a plausible case has to be made that such emotions are indeed felt in the absence of beliefs, since the contexts of imaginative acquaintance are hypothetical. The evidence that such emotions are indeed engendered comes partly from the experience of fiction, which has led philosophers to formulate the "paradox of fictional emotions." Analysis of the premises of the paradox suggests both that these "fictional emotions" are *qualitatively* similar to ordinary, belief-based emotions, and that they do not imply the presence of beliefs required for the emotions in non-fictional contexts.[22] There are also possible cognitive explanations of how the emotions in question could be engendered. Explanations include Gregory Currie's and Ian Ravenscroft's (*Recreative Minds*) theory of belief- and desire-like imaginings; Shaun Nichols' ("Imagining and Believing") single code hypothesis; to a lesser extent (due to its less empirical and more phenomenological emphasis), Kendall Walton's (*Mimesis*) claim that there is a special propositional attitude of "make-belief"; and Damasio's ("Somatic Markers") theory that emotions are activated in hypothetical thought in order to guide practical reasoning. Berys Gaut gives us an illustration of how imaginative acquaintance works when he holds that William Styron's novel, *Sophie's Choice*, acquaints the reader with the phenomenology of being in a situation he has not experienced himself.[23] The experience of Styron's novel suggests that fictions can give us information without which we cannot account properly for either the *consequences* or the motivations of certain moral choices: the novel describes the progressive destruction of an individual by the emotions engendered by her own action, and we cannot make sense of how the emotions cause the moral destruction of the agent unless we have some notion of their phenomenal quality. (Further examples, and explanation of how theories mentioned here are applied to them, are given in section 5.)

---

[22] Radford, "Anna Karenina"; Walton, *Mimesis*, Ch. 7.

[23] Art, Emotion and Ethics, 133–164.

To sum up, certain brain processes cannot be simulated mentally if our brain is not like the target's (for example, if it does not have the same reward pathways), forcing us to admit that theoretical knowledge is sometimes at least jointly required for predicting and understanding the behaviour of target agents. However, there is plausible support for the thesis that mental simulation of emotions and their causal role is possible. With these proposals in hand, we can now examine the role of mental-mental simulation in moral thought experiments.

5. *Moral Thought Experiments*

Now I will draw on the theories described in the preceding section to flesh out, and then to criticise, the thesis that mental simulation can explain the usefulness of moral thought experiments. The problem with the hypothesis, as I see it, is that it presupposes that useful *epistemic*, as opposed to subject-transcendent, discoveries can be made about the nature of values. This bias is explicit when thought experiments are intended to imaginatively acquaint us with the sentimental and emotional circumstances of moral dilemmas. It is less obvious, but equally present, when thought experiments simulate the situation of a decision-making agent. Even simulation understood on the model which accounts least for the emotions, namely Dennett's normatively rational intentional stance, presupposes a set of hierarchical objectives proper to the system ("system objectives"), so that adoption of the intentional stance towards the target system means thinking as if we had its own set of system objectives. Thus, even on a model that does not account for moral sentiments, the simulator will still represent the hypothetical situation *from the target agent's point of view*. But accounting for moral values inevitably means having at some point to account for conflicts *between* sets of system-objectives, and this suggests that the point of view to be adopted should be *external* to the target agent's. "External" here means not so much "allocentric"—since in simulating we adopt another agent's egocentric viewpoint, which means that we are being allocentric—as simply "objective," or "non-epistemic."

Representation of a hypothetical situation from the target agent's point of view is the common denominator of the concepts of "belief-like imaginings" and "desire-like imaginings" in Currie, pretence in Nichols, and make-believe versions of propositional attitudes in Walton: they are claimed to be sufficiently independent of our real beliefs and desires not to conflict with them, so that we can separately assume such attitudes while bracketing our own, putting ourselves in another, or a hypothetical, individual's place. (The same point is made in Leslie's tea-party experiment.)[24] Simulating a decision-making situation adds a number of epistemic features which, according to simulationists, are absent when we merely represent it mentally. One feature is informational: we reproduce the individuality of the situation from the point of view of the putative target agent immersed in it, therefore, we possess more information than we would as an external observer necessarily situated elsewhere. This additional information is represented mentally by means of imagined beliefs, perceptions, desires, and preferences. A second, and crucial, feature is phenomenological: the beliefs, perceptions or desires should normally (for the simulation to comply with the simulation theory of mind) not be represented *de dicto* but *de re*: it is not a case of "I imagine that X believes that *p*," but a case of "I imagine (of myself) that I believe that *p*." In perceptual imaginings such as visualisations, the *re re* form can also have a simple formal object: "I imagine of myself that I see *x*." A third feature is acquaintance with the subjective quality of emotions as outlined in the previous section. It is distinct from the second feature because it concerns emotions as distinct from propositional attitudes, and because a simulation theory could include the first two features while excluding the third. These internal-viewpoint descriptions of moral dilemmas, involving hypothetical adoption of the agent's beliefs and preferences and potentially imaginative acquaintance with feelings, sentiments, and emotions, may be contrasted with external-viewpoint descriptions of moral dilemmas, such as those utilitarians are supposed to use, which are not descriptions from *any* agent's point of view.

Such being the nature of the simulations in question, I turn now to my criticisms of the way they include epistemic features in moral thought experiments. If my criticisms are valid, then either moral thought experiments are valid because they are *not* simulations, or else they are invalid (unreliable for reaching theoretical propositions in ethics) because they *are* simulations.

The mental simulation hypothesis for moral thought experiments would say that process-driven simulations imaginatively acquaint us with epistemic states which are relevant to moral deliberation. This raises two problems: (1) *If* we accept that such epistemic information is indeed

[24] Leslie, "Pretending and Believing."

useful for knowing values, what guarantees that any epistemic information we do so obtain is correct? (2) What justifies the assumption that such epistemic information *is* indeed useful for knowing values? The problems are closely connected; for simplicity, I will not always distinguish them in what follows.

Consider the following case of prediction failure, reported by Stich and Nichols.[25] Asked what a subject would do if it were asked to choose between two apparently similar objects, one to the subject's left, the other to the right, I will predict that they have equal chances of taking either. In fact however, *in vivo*, subjects display a greater propensity to choose the object to the right, so my prediction about the hypothetical situation fails. The first problem this case raises for the hypothesis is, how do we know that simulations of moral dilemmas will not fail to accurately predict the processes in the subject that contribute to his choice, just as they do in the preceding case? This problem is aggravated by the fact that the kinds of philosophical propositions moral though experiments seek to formulate will not be verifiable independently (in the way the left-right choice is) and will not be very numerous, leaving us with hardly any scope for testing them.

The second, and related, problem is one that opposes rationalists and sentimentalists about moral value. Cases of prediction failure favour rationalists. When I attempt to simulate the left-right choice, rationality intrudes and overrides the unconscious processes which *would* make me choose the object to the right *in vivo*: since I am told that the objects are identical, I will *rationally* infer that the choices are equivalent, and (presupposing that the target subject is rational) will attribute to the target hypothetical subject a random choice. In real as opposed to hypothetical circumstances, the rational procedure will not override the unconscious processes which led to the irrational decision. Nevertheless, it remains that *the rational procedure is the one that reaches the right prediction from a normative point of view, even if it is inaccurate from a descriptive point of view*. Agents should choose randomly, even if they are so constituted that they do not choose randomly. According to the imaginative acquaintance thesis, simulation plays the role of adding processes *other* than inference to decision-making; but if a mental simulation succeeded in just this respect, what would guarantee that the factors it introduced were not illegitimate for moral deliberation?

---

[25] Stich and Nichols, "Folk Psychology."

Consider the use of simulation and imaginative acquaintance in Bernard Williams' familiar "Jim and the Indians" thought experiment. Williams' example has been used by Berys Gaut to defend the usefulness of thought experiments—construed as mental simulations and means of imaginative acquaintance—for philosophical purposes.[26] If we bracketed for the moment the detailed information given in the imaginary scenario of Williams' thought experiment, we would state the bare structure of its moral dilemma as follows: all things being equal, should we choose to kill one innocent individual even though we do not want to, or to let twenty innocent individuals be killed by someone else? For a consequentialist, there is more than one way to reach the decision, but all the ways should lead to the former choice. Williams' purpose is to refute consequentialism by bringing out factors which are relevant to the moral decision but which cannot be accounted for by consequentialism. To bring out these factors, he formulates the dilemma by using the following thought experiment:

> Jim finds himself in the central square of a small South American town. Tied up against the wall are a row of twenty Indians, most terrified, a few defiant, in front of them several armed men in uniform. A heavy man in a sweat-stained khaki shirt turns out to be the captain in charge and, after a good deal of questioning of Jim, which establishes that he got there by accident while on a botanical expedition, explains that the Indians are a random group of the inhabitants who, after recent acts of protest against the government, are just about to be killed to remind other possible protestors of the advantages of not protesting. However, since Jim is an honoured visitor from another land, the captain is happy to offer him a guest's privilege of killing one of the Indians himself. If Jim accepts, then as a special mark of the occasion, the other Indians will be let off. Of course, if Jim refuses, then there is no special occasion, and Pedro here will do what he was about to do when Jim arrived, and kill them all. Jim, with some desperate recollection of schoolboy fiction, wonders whether if he got hold of a gun, he could hold the captain, Pedro and the rest of the soldiers to threat, but it is quite clear from the set-up that nothing of the sort is going to work: any attempt at that sort of thing will mean that all the Indians will be killed, and himself. The men against the wall and the other villagers understand the situation, and are obviously begging him to accept. What should he do?[27]

The factors Williams has in mind are "moral feelings" of not being able to "live with" what one has done, and the sense that "each of us is specially

[26] *Art, Emotion and Ethics*; see also Gaut, "L'apprentissage."

[27] Williams, "A Critique of Utilitarianism," 98–99.

responsible for what he does, rather than for what other people do."[28] By their very nature, feelings and the sense of selfhood cannot be appreciated without focusing some attention on the subjective states, and this is what Williams' literary fictional narration is intended to do, unlike my bare bones description further up. For Williams there should strictly speaking be no such bare bones description of the dilemma because, precisely, moral feelings and a sense of moral integrity ("each of us is specially responsible for what he does, rather than for what other people do") *do* enter decision-making. The concept of integrity is cut out to distinguish the agent from other agents, so it implies situatedness, a subjective viewpoint, or egocentricity (in the technical, not the evaluative sense), which cannot be conveyed without the fictional narrative and its literary devices.

Moral feelings and damage to the sense of integrity are not mentioned or described in the fiction, but we are put in a position to appreciate them. A simulationist could hold that moral emotions can be rehearsed mentally in the form of quasi-emotions (described in section 3) or fictional emotions (section 4), and that the narrative causes pretend-beliefs[29] and pretend-desires.[30] The sense of selfhood, which is required for the sense of moral integrity, could be conveyed by privileging the agent's subjectivity over the subjectivity of the other individuals implicated in the dilemma. If the fiction is appreciated by causing *de re* thoughts, then the reader of the thought experiment is in fact placed imaginatively in the situation the dilemma describes. For example, the reader may imagine himself shooting a single peasant at the stake. This does in fact appear to be conveyed by the narration: some details are perceptual, and "a heavy man in a sweat-stained khaki shirt *turns out to be* the captain in charge" is a description from Jim's viewpoint, not the narrator's. We are also given insight into the agent that we could only have by *introspection* ("Jim, with some desperate recollection of schoolboy fiction, wonders whether if he got hold of a gun, he could ..."). In analysing simulation theories of physical thought experiments, we have seen that it is tempting to think there is some form of privileged mental access to nature; in moral thought experiments, the special mental access is to other minds, and in particular to what it feels like to be in another's mind. In this case, though, the arsenal of theories described in section 4 lends the thesis considerable credence:

[28] Ibid., 104, 99.
[29] Nichols and Stich, "Cognitive Penetrability"; Nichols, "Imagining and Believing."
[30] Currie, "Desire"; Currie and Ravenscroft, *Recreative Minds*.

we do seem to have ways of knowing "from the inside," though to a certain fallible degree, the mental life of real or hypothetical others.

Now, for Williams' argument to work, he has to show that the consequences (of killing one of the peasants) for the agent's moral integrity and moral feelings cannot be factored into a consequentialist calculation. Otherwise, the consequentialist will avail of them and simply use the thought experiment to add an extra factor into his calculations. To prevent this, Williams argues:

> ... we are partially at least not utilitarians, and cannot regard our moral feelings merely as objects of utilitarian value. Because our moral relation to the world is partly given by such feelings, and by a sense of what we can or cannot 'live with,' to come to regard those feelings from a purely utilitarian point of view, that is to say, as happenings outside one's moral self, is to lose, in the most literal way, one's integrity. At this point utilitarianism alienates one from one's moral feelings; we shall see a little later how, more basically, it alienates one from one's actions as well.[31]

Taken in isolation, part of Williams' argument is sentimentalist: since we have *moral* feelings which conflict with utilitarian moral judgments, we cannot be utilitarians. But this alone does not establish that negative moral feelings resulting from damage to moral integrity *cannot* be factored into consequentialist calculations. So Williams has to rely on further arguments, alluded to at the end of the citation: that the concept itself of an *agent* presupposes that of integrity. The arguments can be joined by using Williams' distinction between internal and external reasons. If Jim acts against his sense of moral integrity, he will be acting on external reasons (which are internal reasons to, among others, the nineteen peasants). That which allows the motivation of action and the concept of agency in Williams' perspective is partly something akin to Dennett's system-preferences which are *proper* to a system. The difference between "selves" comes down to two factors: (a) one system's system-preferences (desires) concern *it*, not some other system; (b) we are constitutively (in terms of the phenomenology of agency, but also biologically) attached to *our* desires. So Williams' argument comes down to saying that we are not utilitarians because we have moral sentiments which conflict with utilitarianism, and because we could not be agents if we never acted on internal reasons.

Since my purpose is to examine the issues raised only in so far as their resolution can be affected by the use of thought experiments in

[31] Williams, "A Critique of Utilitarianism," 103–104.

ethics, I will restrict myself to mentioning what I think are the relevant replies to Williams in that particular respect. In this respect, I will argue that Williams' position is weakened—not strengthened, as may be supposed—by the use of moral thought experiments involving mental simulations. Specifically, Williams' assumption that epistemic information internal to an agent is useful for knowing values can be challenged from two directions: with an ought/is argument, and with a methodological argument.

We saw in the case of prediction failure that the rational procedure is the one that reaches the right prediction from a *normative* point of view, even if it is inaccurate from a *descriptive* point of view. The problem with Williams' use of the thought experiment is that it privileges the descriptive point of view: the mental simulation reproduces processes pertaining to the phenomenology of selfhood and agency, and to the generation of feelings. It is a descriptive device which shows us how we would feel under certain conditions *because we are so constituted*, in a sense not unlike that appealed to by moral sense theorists such as Hutcheson. So the consequentialist could argue that the kind of knowledge the thought experiment provides is knowledge about how we *are*, whereas the issue is to know how we *ought* to be, and what we should do. The consequentialist appears to have normativity on his side in this dispute, whereas Williams has only an appeal to the authority of description. In fact, the consequentialist can have a more complex position than the sentimentalist, for he can subscribe to one theory concerning the way we are constituted, and to another concerning how we should try to become—*just as* we do in a non-moral context, when we say that "the way we are," which makes us choose objects to our right over objects to our left, is not the way we should be, even for our *own* good. On the possibility of becoming a consequentialist of *some* kind, consider the following sequel to Williams' story. Suppose that Jim makes the choice which means that he has to give up his sense of moral intactness and his emotional serenity. There is a clear sense in which his action is commendable not just in terms of the greater good, but as self-sacrifice *for* the greater good. But in virtue of what is his act commendable in the latter sense, if it is not because he opted for the way he should be, as opposed to the way he is? In defending the moral relevance of feelings and the sense of selfhood, is the sentimentalist not asserting that *even if* a choice is somehow invalid, we should nevertheless stick to it, just because that's how we are?

The methodological argument says that apart from privileging description, the thought experiment performs the function of privileging the agent's point of view and is therefore *epistemologically egocentric*. Consequentialism presupposes precisely that this viewpoint be given up for an objective point of view, so that all parties implicated in the dilemma are weighted equally in moral decision-making. In this sense, the thought experiment *qua* simulation or imaginative acquaintance hides the forest, the consequences for other agents, behind a single tree, the agent's subjectivity. In fact, the same thought experiment could also be used to show how an objective assessment can be clouded with epistemic illusions about the objective value of our personal integrity. It seems that to combat consequentialism, one would have to combat precisely its commitment to an objective viewpoint—and this, the mental simulation cannot do, because it operates from *within* an egocentric framework. To borrow terms used by Wilfrid Sellars in a different context, it "operates within a framework, and cannot support that framework."

Such arguments do not apply only to Williams' thought experiments, but suggest more generally that if moral thought experiments are mental simulations, then their use must privilege certain moral theories—sentimentalist ones, for example—over others. Combined with the problem of prediction failure, this should make them unreliable for formulating general propositions about the nature of value. If thought experiments are a reliable method for ethical thought, they must succeed in virtue of some procedure other than mental simulation.

BIBLIOGRAPHY

Adams, Ernest: *The Logic of Conditionals* (Dordrecht, 1975).

Albert of Saxony: *Quaestiones in Aristotelis De caelo*, ed. Benoît Patar (Louvain-la-Neuve, 2008).

Alexander of Aphrodisias: *Quaestiones* in *Alexandri Aphrodisiensis praeter commentaria scripta minora*, ed. Ivo Bruns, Commentaria in Aristotelem Graeca, suppl. 2.2 (Berlin, 1892).

Alexander, H.G. (ed.): *The Leibniz-Clarke Correspondence. With Extracts from Newton's Principia and Opticks* (Manchester, 1956).

Algra, Keimpe: *Concepts of Space in Greek Thought* (Leiden, 1995).

Allison, Henry: *Kant's Transcendental Idealism: An Interpretation and Defense* (New Haven, CT, 2004 [2nd ed.]).

Aquinas, Thomas: *In Aristotelis libros De caelo et mundo, De generatione et corruptione et Meteorologicorum expositio*, ed. Raymundi M. Spiazzi (Turin, 1952).

——, *In octo libros Physicorum Aristotelis expositio*, ed. P. Mariano Maggiòlo (Turin, 1965).

——, *Summa theologiae*, ed. Petri Caramello (Turin, 1948–1950).

Aristotle: *De anima*, ed. David Ross (Oxford, 1961).

——, *De caelo*, ed. Paul Moraux (Paris, 1965).

——, *Metaphysics X–XIV, Œconomica, and Magna moralia*, trans. Hugh Tredennick and G. Cyril Armstrong (Cambridge, MA, 1935).

——, *On the Heavens*, ed. and trans. William K. Chambers Guthrie (Cambridge, MA, 1934).

——, *Physics*, ed. and trans. Philip H. Wickstead and Francis M. Cornford (Cambridge, MA, 1934).

——, *Physics*, ed. David Ross (Oxford, 1936).

——, *Politics*, trans. Harris Rackham (Cambridge, MA, 1932).

Atkinson, David: "Experiments and Thought Experiments in Natural Sciences," in *Observation and Experiment in the Natural and Social Sciences*, ed. Maria Carla Galavotti (Dordrecht, 2003), 209–225.

Atkinson, David and Jeanne Peijnenburg: "Galileo and Prior Philosophy," *Studies in History and Philosophy of Science* 35 (2004), 115–136.

——, "When Are Thought Experiments Poor Ones?," *Journal for General Philosophy of Science* 34 (2003), 305–322.

Audi, Robert (ed.): *The Cambridge Dictionary of Philosophy* (Cambridge, 1999 [2nd ed.]).

Averroes: *Averroes' Questions in Physics*, trans. Helen Tunik Goldstein (Dordrecht, 1991).

——, *Aristotelis opera cum Averrois commentariis* (Venice, 1562–1574).

——, *Jawāmiʿ kitāb al-samāʿ al-ṭabīʿī*, ed. Josep Puig (Madrid, 1983).

——, *Talkhīṣ fī al-samāʾ wa al-ʿālam*, ed. J. al-Dīn al-ʿAlawī (Fez, 1984).

——, *Talkhīṣ kitāb al-qiyās*, ed. M.M. Qāsim et al. (Cairo, 1983).

Avicenna: *Al-Shifāʾ: Al-samāʿ al-ṭabīʿī* ["The Healing: The Physics"], ed. Said Zāyid and Ibrahim Madkūr (Cairo, 1983).

Bailey, Cyril (ed.): *Titi Lucreti Cari De rerum natura libri sex*, vol. 2: Commentary Books I–III (Oxford, 1947).

Barenghi, Giovanni: *Considerazioni del signor Giovanni Barenghi sopra il Dialogo de'duoi massimi sistemi tolemaico, e copernicano del Sig. Accademico Linceo* (Pisa, 1638).

Bealer, George: "Modal Epistemology and the Rationalist Renaissance," in *Conceivability and Possibility*, ed. Tamar Szabó Gendler and John Hawthorne (Oxford, 2002).

Bertoloni Meli, Domenico: "The Role of Numerical Tables in Galileo and Mersenne," *Perspectives on Science* 12 (2004), 164–190.

Bianchi, Luca: *Il vescovo e i filosofi. La condanna parigina del 1277 e l'evoluzione dell'aristotelismo scolastico* (Bergamo, 1990).

Bignone, Ettore: *Storia della Letteratura Latina*, vol. 2 (Florence, 1945).

Bonjour, Laurence: *In Defense of Pure Reason* (Cambridge, 1996).

Bossier, Fernand and Carlos Steel: "Priscianus Lydus en de 'In De anima' van Pseudo(?)-Simplicius," *Tijdschrift voor filosofie* 34 (1972), 761–822.

Bouveresse, Jacques: *La connaissance de l'écrivain* (Marseille, 2008).

Bowin, John: "Chrysippus' Puzzle About Identity," *Oxford Studies in Ancient Philosophy* 24 (2003), 239–251.

Brisson, Luc: "The Role of Myth in Plato and its Prolongations in Antiquity," *The European Legacy* 12 (2007), 141–158.

Brown, James Robert: *Laboratory of the Mind: Thought Experiments in the Natural Sciences* (London, 1991).

——, "Thought Experiments: a Platonic Account," in *Thought Experiments in Science and Philosophy*, ed. Tamara Horowitz and Gerald J. Massey (Savage, MD, 1991), 119–128.

——, "Thought Experiments," in *A Companion to the Philosophy of Science: Studies in Honor of Wendell Clausen*, ed. William H. Newton-Smith (Oxford, 2000), 528–531.

——, "Thought Experiments," *The Stanford Encyclopedia of Philosophy*, ed. Edward N. Zalta (2007). ⟨http://plato.stanford.edu/archives/fall2009/entries/thought-experiment/⟩

——, "Why Thought Experiments Transcend Empiricism," in *Contemporary Debates in the Philosophy of Science*, ed. Christopher Hitchcock (Oxford, 2004), 23–43.

Buonamici, Francesco: *De motu libri X* (Florence, 1591).

Buridan, John: *Expositio et quæstiones in Aristotelis De cælo*, ed. Benoît Patar (Louvain-la-Neuve, 1996).

——, *Quaestiones de Anima (tertia sive ultima lectura)*, in *John Buridan's Philosophy of Mind*, ed. and trans. Jack Zupko (Ithaca, NY, 1989).

——, *Quaestiones super octos libros Physicorum (tertia lectura)*, unpublished transcription by Jean Celeyrette.

——, *Quaestiones super octos libros Physicorum (ultima lectura)* (Frankfurt am Main, 1964 [reproduction of the 1509 printing]).

Byrne, Ruth: *The Rational Imagination* (Cambridge, MA, 2005).

Carruthers, Peter and Peter Smith (eds.): *Theories of Theories of Mind* (Cambridge, 1996), 119–137.
Cartwright, Nancy: "The Vanity of Rigor in Economics: Theoretical Models and Galilean Experiments," in *Hunting Causes and Using Them. Approaches in Philosophy and Economics* (Cambridge, 2007), 217–236.
Celeyrette, Jean: "La problématique du point chez Jean Buridan," *Vivarium* 42 / 1 (2004), 86–108.
Chatton, Walter: *Reportatio super sententias*, Liber II, ed. Joseph C. Wey and Girard J. Etzkorn (Toronto, 2004).
Cicero: *De officiis*, ed. Michael Winterbottom (Oxford, 1994).
———, *De officiis libri III*, ed. Olof Gigon (Zurich, 1950).
———, *De re publica*, ed. Konrat Ziegler (Stuttgart, 1969).
Clagett, Marshall: *The Science of Mechanics in the Middle Ages* (Madison, WI, 1959).
Cleomedes: *Caelestia*, ed. Robert B. Todd (Leipzig, 1990).
Cohen, Martin: *Wittgenstein's Beetle and Other Classic Thought Experiments* (Oxford, 2005).
Cohnitz, Daniel: "Ørsteds 'Gedankenexperiment': eine Kantianische Fundierung der Infinitesimalrechnung? Ein Beitrag zur Begriffsgeschichte von 'Gedankenexperiment' und zur Mathematikgeschichte des frühen 19. Jahrhunderts," *Kant-Studien* 99 / 4 (2008), 407–433.
Currie, Gregory: "Mental Simulation and Motor Imagery," *Philosophy of Science* 64 (1997), 161–180.
———, *The Nature of Fiction* (Cambridge, MA, 1990).
———, "Simulation-Theory, Theory-Theory and the Evidence from Autism," in *Theories of Theories of Mind*, ed. Peter Carruthers and Peter Smith (Cambridge, 1996), 242–256.
———, "Visual Imagery as the Simulation of Vision," *Mind & Language* 10 (1995), 25–44.
———, "Desire in Imagination," in *Conceivability and Possibility*, ed. Tamar Szabó Gendler and John Hawthorne (Oxford, 2002), 201–221.
Currie, Gregory and Ian Ravenscroft: *Recreative Minds* (Oxford, 2003).
Dales, Richard: "Henry of Harclay on the Infinite," *Journal of the History of Ideas* 45 (1984), 295–301.
Damasio, Antonio, Daniel Tranel and Hanna Damasio: "Somatic Markers and the Guidance of Behavior: Theory and Preliminary Testing," in *Frontal Lobe Function and Dysfunction*, ed. H.S. Levin, H.M. Eisenberg and A.L. Benton (New York, 1991), 217–229.
Davidson, Donald: *Essays on Actions and Events* (New York, 1980).
Davies, Martin: "The Mental Simulation Debate," in *Objectivity, Simulation and the Unity of Consciousness: Current Issues in the Philosophy of Mind. Proceedings of the British Academy 83*, ed. Christopher Peacocke (Oxford, 1994), 99–127.
Davies, Martin and Tony Stone: "Folk Psychology and Mental Simulation," in *Contemporary Issues in the Philosophy of Mind*, ed. Anthony O'Hear (Cambridge, 1998), 53–82.
———, "Simulation Theory," *Routledge Encyclopedia of Philosophy* (London, 2000).

de Sousa, Ronald: *The Rationality of Emotion* (Cambridge, MA, 1987).
Dennett, Daniel: *The Intentional Stance* (Cambridge, MA, 1989).
Descartes, René: "Meditations on the First Philosophy," in *The Philosophical Works of Descartes*, trans. Elizabeth Haldane and G.R.T. Ross (Cambridge, 1981), I: 131–200.
——, *Œuvres de Descartes*, ed. Charles Adam and Paul Tannery, new pres. by Bernard Rochot and Pierre Costabel, 11 vols. (Paris, 1964–1974).
Dretske, Fred: *Knowledge and the Flow of Information* (Cambridge, MA, 1981/ Stanford, 1999).
Duhem, Pierre: *Études sur Léonard de Vinci* (Paris, [1906–1913] 1955).
——, *Medieval Cosmology, Theories of Infinity, Place, Time, Void and the Plurality of Worlds*, ed. and trans. Roger Ariew (Chicago, 1985).
——, *La théorie physique* (Paris, [1906] 1981).
Duns Scotus, John: *De primo principio*, ed. Wolfgang Kluxen (Darmstadt, 1974).
Einstein, Albert: *Autobiographical Notes: A Centennial Edition*, trans. and ed. Paul Arthur Schilpp (La Salle, IL, 1979), 1–94.
——, *The Collected Papers of Albert Einstein. Vol. 8: The Berlin Years: Correspondence, 1914–1918 (Parts A and B)*, ed. Robert Schulmann, A.J. Kox, Michel Janssen and József Illy (Princeton, NJ, 1998).
——, "Discussions of Lectures in Bad Nauheim," in *The Collected Papers of Albert Einstein. Vol. 7: The Berlin Years: Writings, 1918–1921*, ed. Michel Janssen, Robert Schulmann, József Illy, Christoph Lehner and Diana Kormos Buchwald (Princeton, NJ, 2002), 351–359.
——, "The Foundation of the General Theory of Relativity," in *The Collected Papers of Albert Einstein. Vol. 6: The Berlin Years: Writings, 1914–1917*, ed. Martin Klein, A.J. Kox and Robert Schulmann (Princeton, NJ, 1996), 146–200.
——, *Letters to Solovine*, trans. Wade Baskin (New York, 1987).
——, "On the Electrodynamics of Moving Bodies," in *The Collected Papers of Albert Einstein. Vol. 2: The Swiss Years: Writings, 1900–1909*, ed. John Stachel, Robert Schulmann, David C. Cassidy and Jürgen Renn (Princeton, NJ, 1989), 140–171.
——, "On the Influence of Gravitation on the Propagation of Light," in *The Collected Papers of Albert Einstein. Vol. 3: The Swiss Years: Writings, 1909–1911*, ed. Martin Klein, A.J. Kox, Jürgen Renn and Robert Schulmann (Princeton, NJ, 1993), 379–387.
——, "On the Present State of the Problem of Gravitation," in *The Collected Papers of Albert Einstein. Vol. 4: The Swiss Years: Writings, 1912–1914*. ed. Martin Klein, A.J. Kox, Jürgen Renn and Robert Schulmann (Princeton, NJ, 1995), 198–222.
——, "On the Special and the General Theory of Relativity (A Popular Account)," in *The Collected Papers of Albert Einstein. Vol. 6: The Berlin Years: Writings, 1914–1917*, ed. Martin Klein, A.J. Kox and Robert Schulmann (Princeton, NJ, 1996), 247–420.
——, "On the Theory of Tetrode and Sackur for the Entropy Constant," in *The Collected Papers of Albert Einstein. Vol. 6: The Berlin Years: Writings, 1914–1917*, ed. Martin Klein, A.J. Kox and Robert Schulmann (Princeton, NJ, 1996), 121–131.

——, "Time, Space and Gravitation," in *The Collected Papers of Albert Einstein. Vol. 7: The Berlin Years: Writings, 1918–1921*, ed. Michel Janssen, Robert Schulmann, József Illy, Christoph Lehner and Diana Kormos Buchwald (Princeton, NJ, 2002), 213–215.
Einstein, Albert and Leopold Infeld: *The Evolution of Physics* (New York, 1961).
Engel, Pascal: "Des avantages et des inconvénients de faire de la philosophie analytique en fauteuil," in *La philosophie*, ed. Melika Ouelbani (Tunis, 2006).
——, "Is there a Geography of Thought?," *Cognitio. Revista de Filosofia* 8 (2007), 197–212.
Epicurus: *Epistula ad Herodotum* in *Epicuro. Opere*, ed. Graziano Arrighetti (Turin, 1973).
Fine, Arthur: *The Shaky Games: Einstein, Realism and the Quantum Theory* (Chicago, 1986).
Foot, Philippa: "The Problem of Abortion and the Doctrine of the Double Effect," reprinted in *Virtues and Vices* (Oxford, 1978), 19–32.
Funkenstein, Amos: *Theology and the Scientific Imagination* (Princeton, 1986).
Galilei, Galileo: *Dialogo sopra i due massimi sistemi del mondo tolemaico e copernicano*, ed. Ottavio Besomi and Mario Helbing (Padua, 1998).
——, *Dialogue Concerning the Two Chief World Systems*, ed. and trans. Stillman Drake (Berkeley, 1967).
——, *On Motion and On Mechanics: Comprising De Motu* (ca. 1590), trans. with Introduction and Notes by I.E. Drabkin and *Le Meccaniche* (ca. 1600), trans. with Introduction and Notes by Stillman Drake (Madison, 1960).
——, *Two New Sciences*, ed. and trans. Stillman Drake (Toronto, 1989).
Gardner, Sebastian: *Kant and the Critique of Pure Reason* (Abingdon and New York, 1999).
Gaut, Berys: "L'apprentissage éthique, l'art et l'imagination," in *Ce que l'art nous apprend* (Nancy, 2008), 23–35.
——, *Art, Emotion and Ethics* (Oxford, 2007).
Gendler, Tamar Szabó: "Galileo and the Indispensability of Scientific Thought Experiment," *The British Journal for the Philosophy of Science* 49 (1998), 397–424.
——, *Imaginary Exceptions: On the Power and Limits of Thought Experiments* (New York, 2000).
——, "Thought Experiments Rethought—and Reperceived," *The British Journal for the Philosophy of Science* 71 (2004), 1152–1163.
Gendler, Tamar Szabó and John Hawthorne (eds.): *Conceivability and Possibility* (Oxford, 2002).
Gendler, Tamar Szabó and Karson Kovakovich: "Genuine Rational Fictional Emotions," in *Contemporary Debates in Aesthetics*, ed. Matthew Kieran (Oxford, 2005), 241–253.
Gettier, Edmund: "Is Justified True Belief Knowledge?," *Analysis* 23 (1963), 121–123.
Glasner, Ruth: "Ibn Rushd's Theory of *minima naturalia*," *Arabic Sciences and Philosophy* 11 (2001), 9–26.
Glock, Hans-Johann (ed.): *Strawson and Kant* (Oxford, 2003).

Goldman, Alvin I.: "In Defense of the Simulation Theory," *Mind & Language* 7 (1992), 104–119.
Gooding, David: "What is *Experimental* about Thought Experiments?," *Proceedings of the Biennial Meeting of the Philosophy of Science Association 1992*, vol. 2, ed. David Hull, Micky Forbes and Kathleen Okruhlik (East Lansing, MI, 1993), 280–290.
Gordon, Robert M.: "The Simulation Theory: Objections and Misconceptions," *Mind and Language* 7 (1992), 11–34.
Grabmann, Martin: *Guglielmo di Moerbeke O.P., il traduttore delle opere di Aristotele* (Rome, 1946).
Grant, Edward: "The Condemnation of 1277, God's Absolute Power, and Physical Thought in the Late Middle Ages," *Viator* 10 (1979), 211–244.
——, *God and Reason in the Middle Ages* (Cambridge, 2001).
——, "Medieval and Seventeenth-century Conceptions of an Infinite Void Space beyond the Cosmos," *Isis* 60 (1969), 39–60.
——, "Medieval Natural Philosophy: Empiricism without Observation," in *The Dynamics of Aristotelian Natural Philosophy from Antiquity to the Seventeenth Century*, ed. Cees Leijenhorst, Christoph Lüthy and Hans Thijssen (Leiden, 2002), 141–168.
——, *Much Ado about Nothing: Theories of Space and Vacuum from the Middle Ages to the Scientific Revolution* (Cambridge, 1981).
——, *A Source Book in Medieval Science* (Cambridge, MA, 1974).
Hacker, Peter: "A Philosopher of Philosophy," *Philosophical Quarterly* 59/235 (2009), 337–348.
Hadot, Ilsetraut: "Simplicius or Priscianus? On the Author of the Commentary on Aristotle's *De Anima* (CAG XI): A Methodological Study," *Mnemosyne* 55 (2002), 159–199.
Häggqvist, Soren: *Thought Experiments in Philosophy* (Stockholm, 1996).
Hahm, David E.: *The Origins of Stoic Cosmology* (Columbus, OH, 1977).
Heal, Jane: "Understanding Other Minds from the Inside," in *Contemporary Issues in the Philosophy of Mind*, ed. Anthony O'Hear (Cambridge, 1998), 83–99.
Hitchcock, Christopher (ed.): *Contemporary Debates in the Philosophy of Science* (Oxford, 2004).
Holton, Gerald: "Mach, Einstein, and the Search for Reality," in *Ernst Mach. Physicist and Philosopher*, ed. Robert S. Cohen and Raymond J. Seeger (Dordrecht, 1970), 165–199.
Horowitz, Tamara and Gerald J. Massey (eds.): *Thought Experiments in Science and Philosophy* (Savage, MD, 1991).
Huffman, Carl A.: *Archytas of Tarentum. Pythagorean, Philosopher and Mathematician King* (Cambridge, 2005).
Hugonnard-Roche, Henri: "L'hypothétique et la nature dans la physique parisienne du XIVe siècle," in *La nouvelle physique du XIVe siècle*, ed. Stefano Caroti and Pierre Souffrin (Florence, 1997), 161–177.
——, "Modalités et argumentation chez Nicole Oresme," in *Nicole Oresme*, ed. Alain Segonds and Pierre Souffrin (Paris, 1988), 145–163.
Hume, David: "An Enquiry Concerning Human Understanding," in *Enquiries*

Concerning Human Understanding and Concerning the Principles of Morals, ed. Peter H. Nidditch (Oxford, 1975 [3rd ed.]).
——, *A Treatise of Human Nature*, ed. Peter H. Nidditch (Oxford, 1980 [2nd ed.]).
Ierodiakonou, Katerina: "Ancient Thought Experiments: A First Approach," *Ancient Philosophy* 25 (2005), 125–141.
Jamieson, Dale: "Method and Moral Theory," in *A Companion to Ethics*, ed. Peter Singer (Oxford, 1991), 476–490.
Janis, Allen J.: "Can Thought Experiments Fail?," in *Thought Experiments in Science and Philosophy*, ed. Tamara Horowitz and Gerald J. Massey (Savage, MD, 1991), 113–118.
Johnson-Laird, Philip: "Mental Models," *Foundations of Cognitive Science*, ed. Michael Posner (Cambridge, MA, 1989), 469–499.
Kant, Immanuel: *Critique of Practical Reason*, in *Practical Philosophy*, trans. and ed. Mary Gregor (Cambridge, 1996), 153–271.
——, *Critique of Pure Reason*, ed. and trans. Paul Guyer and Allan W. Wood (Cambridge, 1998).
——, *De mundi sensibilis atque intelligibilis forma et principiis*, in *Kants Werke*, Akademie Textausgabe (Berlin, 1968), II: 385–420.
——, *Prolegomena zu einer jeden künftigen Metaphysik, die als Wissenschaft wird auftreten können*, in *Kants Werke*, Akademie Textausgabe (Berlin, 1968), IV: 253–384.
——, "Von dem ersten Grunde des Unterschiedes der Gegenden im Raume," in *Kants Werke*, Akademie Textausgabe (Berlin, 1968), II: 375–384.
King, Peter: "Mediaeval Thought-Experiments: The Metamethodology of Mediaeval Science," in *Thought Experiments in Science and Philosophy*, ed. Tamara Horowitz and Gerald J. Massey (Savage, MD, 1991), 43–64.
Knuuttila, Simo: "Medieval Modal Theories and Modal Logic," in *Handbook of the History of Logic* 2, ed. Dov Gabbay and John Woods (Amsterdam, 2008), 505–578.
——, "Modality as Alternativeness in John Duns Scotus," in *New Essays on Metaphysics as Scientia Transcendens. Textes et études du Moyen Âge 43*, ed. Roberto Hofmeister Pich (Louvain-la-Neuve, 2007), 145–157.
——, "Necessities in Buridan's Natural Philosophy," in *The Metaphysics and Natural Philosophy of John Buridan, Medieval and Early Modern Science* 2, ed. Hans Thijssen and Jack Zupko (Leiden, 2001), 65–76.
Kornblith, Hilary: "Timothy Williamson's *The Philosophy of Philosophy*," *Analysis* 69/1 (2009), 109–116.
Koyré, Alexandre: "Le *De Motu gravium* de Galilée. De l'expérience imaginaire et de son abus," reprinted in *Études d'histoire de la pensée scientifique* (Paris, 1973), 224–271.
——, "Pascal savant," reprinted in *Études d'histoire de la pensée scientifique* (Paris, 1973), 362–389.
Kretzmann, Norman: "Adam Wodeham's Anti-Aristotelian Anti-Atomism," *History of Philosophy Quarterly* 1 (1984), 381–397.
Kuhn, Thomas: "A Function for Thought Experiment," in *The Essential Tension* (Chicago, 1977), 240–265.

Kühne, Ulrich: *Die Methode des Gedankenexperiments* (Frankfurt am Main, 2005).

Kukkonen, Taneli: "Alternatives to Alternatives: Approaches to Aristotle's Arguments *per impossibile*," *Vivarium* 40 (2002), 137–173.

——, "The Impossible, insofar as It Is Possible: Ibn Rushd and Jean Buridan on Logic and Natural Theology," in *Logik und Theologie. Das Organon im arabischen und im lateinischen Mittelalter*, ed. Dominik Perler and Ulrich Rudolph (Leiden, 2005), 447–467.

——, "Mind and Modal Judgement: Al-Ghazālī and Ibn Rushd on Conceivability and Possibility," in *Mind and Modality*, ed. Vesa Hirvonen, Toivo J. Holopainen and Miira Tuominen (Leiden, 2006), 121–139.

Lactantius: *Epitome divinarum institutionum*, ed. Eberhard Heck and Antonie Wlosok (Stuttgart/Leipzig, 1994).

Lagerlund, Henrik: *Modal Syllogistics in the Middle Ages* (Leiden, 2000).

Lautner, Peter: "The Status and Method of Psychology According to the Late Neoplatonists and their Influence during the Sixteenth Century," in *The Dynamics of Aristotelian Natural Philosophy from Antiquity to the Seventeenth Century*, ed. Cees Leijenhorst, Christoph Lüthy and Hans Thijssen (Leiden, 2002), 81–109.

Le Guin, Ursula K.: "Introduction to *The Left Hand of Darkness*," reprinted in *The Language of the Night: Essays on Fantasy and Science Fiction* (New York, 1979), 155–159.

——, "Is Gender Necessary?," reprinted in *The Language of the Night: Essays on Fantasy and Science Fiction* (New York, 1979), 161–169.

Lehrer, Keith: *Theory of Knowledge* (Boulder, 1990).

Leslie, Alan: "Pretending and Believing: Issues in the Theory of ToMM," *Cognition* 50 (1994), 211–238.

Lewis, David: "Truth in Fiction," in *Philosophical Papers* (Oxford, 1975).

——, *Counterfactuals* (Oxford, 1973).

Locke, John: *An Essay Concerning Human Understanding*, ed. Peter H. Nidditch (Oxford, 1979).

Lucretius: *De rerum natura*, ed. Martin Ferguson Smith (London/Cambridge, MA, 1975).

——, *On the Nature of Things*, trans. Martin Ferguson Smith (Indianapolis, IN, 2001).

Lyons, Malcom Cameron: *The Arabic Version of Themistius' "De anima"* (Oxford, 1973).

Mach, Ernst: *Knowledge and Error. Sketches on the Psychology of Enquiry*, trans. Thomas J. McCormack and Paul Foulkes (Dordrecht, 1976).

——, *The Science of Mechanics*, trans. Thomas J. McCormack (La Salle, IL, 1960).

Martin, Christopher: "Non-reductive Arguments from Impossible Hypotheses in Boethius and Philoponus," *Oxford Studies in Ancient Philosophy* 17 (1999), 279–302.

McAllister, James: "The Evidential Significance of Thought Experiment in Science," *Studies in History and Philosophy of Science* 27 (1996), 233–250.

——, "The Virtual Laboratory: Thought Experiments in Seventeenth-Century

Mechanics," in *Collection, Laboratory, Theater: Scenes of Knowledge in the 17th Century*, ed. Helmar Schramm, Ludger Schwarte and Jan Lazardzig (New York, 2005), 35–56.
——, "Thought Experiments and the Belief in Phenomena," *Philosophy of Science* 71 (2004), 1164–1175.
Milmed, Bella: "Possible Experience and Recent Interpretations of Kant," in *Kant's Studies Today*, ed. Lewis W. Beck (Lasalle, 1969), 301–321.
Mulligan, Kevin: "From Appropriate Emotions to Values," *Monist: Secondary Qualities Generalised* (1998), 161–188.
Murdoch, John E.: "The Analytic Character of Late Medieval Learning. Natural Philosophy without Nature," in *Approaches to Nature in the Middle Ages*, ed. Lawrence Roberts (New York, 1982), 171–213.
——, "Henry of Harclay and the Infinite," *Studi sul XIV secolo in memoria di Anneliese Maier*, ed. Alfonso Maierù (Rome, 1981), 220–261.
Needham, Amy and Renée Baillargeon: "Intuitions about Support in 4.5-month-old Infants," *Cognition* 47 (1993), 121–148.
Nersessian, Nancy: "The Cognitive Basis of Model-based Reasoning in Science," in *The Cognitive Basis of Science*, ed. Peter Carruthers, Stephen Stich and Michael Siegal (Cambridge, 2002), 133–153.
——, "In the Theoretician's Laboratory: Thought Experimenting as Mental Modeling," *Proceedings of the Biennial Meeting of the Philosophy of Science Association* (1992), II: 291–301.
——, "Model-based Reasoning in Conceptual Change," in *Model-Based Reasoning in Scientific Discovery*, ed. Lorenzo Magnani, Nancy Nersessian and Paul Thagard (New York, 1999), 5–22.
Nicholas of Autrecourt, *Universal Treatise [Exigit ordo]*, trans. Leonard Kennedy et al. (Milwaukee, WI, 1971).
Nichols, Shaun: "Imagining and Believing: the Promise of a Single Code," *The Journal of Aesthetics and Art Criticism* 62 / 2 (2004), 129–139.
Nichols, Shaun and Stephen Stich: "Cognitive Penetrability, Rationality and Restricted Simulation," *Mind & Language* 12 (1997), 297–326.
Nichols, Shaun, Stephen Stich and Jonathan Weingart: "Meta-Skepticism," in *The Skeptics*, ed. Stephen Luper (Aldershot, 2003), 227–247.
Nisbett, Richard E.: *The Geography of Thought: How Asians and Westerners Think Differently … and Why* (New York, 2003).
North, John D.: *Stars, Minds, and Fate: Essays in Ancient and Medieval Cosmology* (Cambridge, 1979).
Norton, John D.: "Are Thought Experiments just What You Thought?," *Canadian Journal of Philosophy* 26 (1996), 333–366.
——, "On Thought Experiments: Is there more to the Argument?," *Philosophy of Science* 71 (2004), 1139–1151.
——, "Thought Experiments in Einstein's Work," in *Thought Experiments in Science and Philosophy*, ed. Tamara Horowitz and Gerald J. Massey (Savage, MD, 1991), 129–148.
——, "Why Thought Experiments Do not Transcend Empiricism," in *Contemporary Debates in the Philosophy of Science*, ed. Christopher Hitchcock (Oxford, 2004), 44–66.

Oresme, Nicholas: *Le livre du ciel et du monde*, ed. and trans. Albert D. Menut and Alexander J. Denomy (Madison, WI, 1968).
———, *Quaestiones de Spera*, ed. and trans. Garrett Droppers, unpublished PhD dissertation (University of Wisconsin, 1966).
———, *Quaestiones super octos libros Physicorum*, ed. Jean Celeyrette and Edmond Mazet (forthcoming).
Ørsted, Hans Christian: "First Introduction to General Physics," in *Selected Scientific Works* (Princeton, [1811] 1998), 282–309.
———, *Selected Scientific Works*, trans. and ed. Karen Jelved, Andrew D. Jackson and Ole Knudsen with an introduction by Andrew D. Wilson (Princeton, NJ, 1998).
Pabst Battin, Margaret: *Ethical Issues in Suicide* (Prentice-Hall, NJ, 1995).
Peacocke, Christopher: *The Realm of Reasons* (Oxford, 2004).
Perkams, Matthias: "Priscian of Lydia, Commentator on the *De Anima* in the Tradition of Iamblichus," *Mnemosyne* 58 (2005), 510–530.
Philo: *De aeternitate mundi* in *Philonis Alexandrini opera quae supersunt*, 7 vols., ed. Leopold Cohn and Siegfried Reiter (Berlin, 1915–1930).
Philoponus: *In Aristotelis Physicorum libros octo commentaria*, 2 vols., ed. Hieronymus Vitelli, Commentaria in Aristotelem Graeca 16 & 17 (Berlin, 1887 and 1888).
———, *In Aristotelis De anima libros commentaria*, ed. Michael Hayduck, Commentaria in Aristotelem Graeca 15 (Berlin, 1897).
———, *Corollaries on Place and Void*, trans. David Furley (London, 1991).
———, *On Aristotle On the Soul 3.1–8*, trans. William Charlton (London/Ithaca, NY, 2000).
———, *On Aristotle On the Soul 3.9–13 with Stephanus, On Aristotle On Interpretation*, trans. William Charlton (London/Ithaca, NY, 2000).
Plato: *Timaeus*, in *Platonis Opera*, vol. 4, ed. John Burnet (Oxford, 1963).
Plotinus: *Enneades IV–V*, in *Opera*, vol. 2 (Oxford, 1977).
Plutarch: *Theseus*, in *Plutarchi vitae parallelae*, vol. 1.1, ed. Konrat Ziegler (Leipzig, 1969).
Popper, Karl: *The Logic of Scientific Discovery* (London, 2002).
Posidonius: *The Fragments*, ed. Ludwig Edelstein and Ian G. Kidd (Cambridge, 1989).
Prinz, Jesse J.: *Furnishing the Mind: Concepts and Their Perceptual Basis* (Cambridge, MA, 2002).
Priscian of Lydia: *Metaphrasis in Theophrastum*, ed. Ingram Bywater (Berlin, 1886).
———, *On Theophrastus On Sense Perception*, trans. Pamela Huby, publ. with *Simplicius, On Aristotle On the Soul* 2.5–12, trans. Carlos Steel, notes Peter Lautner (London, 1997).
Proclus: *Procli philosophi Platonici opera inedita*, ed. Victor Cousin (Paris, 1864).
Radford, Colin: "How Can We Be Moved by the Fate of Anna Karenina?," *Proceedings of the Aristotelian Society*, suppl. vol. 49 (1975), 67–80.
Reisberg, Daniel: "The Non-ambiguity of Mental Images," in *Stretching the Imagination: Representation and Transformation in Mental Imagery*, ed. Cesare Cornoldi et al. (New York, 1996), 127–131.

Rescher, Nicholas: "Thought Experimentation in Pre-Socratic Philosophy," in *Thought Experiments in Science and Philosophy*, ed. Tamara Horowitz and Gerald J. Massey (Savage, MD, 1991), 31–41.
——, *What if? Thought Experimentation in Philosophy* (New Brunswick, 2005).
Rijen, Jeroen van: *Aspects of Aristotle's Logic of Modalities* (Dordrecht, 1989).
Schildknecht, Christiane: *Philosophische Masken: Literarische Formen der Philosophie bei Platon, Descartes, Wolff und Lichtenberg* (Stuttgart, 1990).
Sedley, David: "The Stoic Criterion of Identity," *Phronesis* 27 (1982), 255–275.
Simplicius: *In Aristotelis De caelo commentaria*, ed. Johann Ludwig Heiberg, Commentaria in Aristotelem Graeca 7 (Berlin, 1894).
——, *In Aristotelis Physicorum libros octo commentaria*, 2 vols., ed. Hermann Diels, Commentaria in Aristotelem Graeca 9 and 10 (Berlin, 1882 and 1895).
——, *In libros Aristotelis De anima commentaria*, ed. Michael Hayduck, Commentaria in Aristotelem Graeca 11 (Berlin, 1882).
——, *On Aristotle On the Heavens 1.5–9*, trans. Robert James Hankinson (London, 2004).
Sirridge, Mary: '"The Living Thing is either Nothing or Posterior'. Radulphus Brito's *Quaestiones super libros* De anima," in *Mind, Cognition and Representation. The Tradition of Commentaries on Aristotle's* De anima, ed. Paul Bakker and Hans Thijssen (Aldershot, 2007), 45–68.
Slater, Alan, Victoria Morison, Marcia Somers, Anne Mattock, Elizabeth Brown and David Taylor: "Newborn and Older Infants' Perception of Partly Occluded Objects," *Infant Behavior and Development* 13 (1990), 33–49.
Smart, John Jamieson Carswell and Bernard Williams: *Utilitarianism: For and Against* (Cambridge, 1973).
Sorabji, Richard: *Matter, Space & Motion: Theories in Antiquity and their Sequel* (London, 1988).
——, *Time, Creation and the Continuum: Theories in Antiquity and the Early Middle Ages* (London, 1983).
Sorensen, Roy: *Thought Experiments* (Oxford, 1992).
——, "Thought Experiments and the Epistemology of Laws," *Canadian Journal of Philosophy* 22 / 1 (1992), 15–44.
Sosa, Ernesto: "A Defense of the Use of Intuitions in Philosophy," in *Stich and His Critics*, ed. Dominic Murphy and Michael Bishop (Oxford, 2009).
Stalnaker, Robert: "A Theory of Conditionals," in *Studies in Logical Theory*, ed. Nicholas Rescher (Oxford, 1968), 98–112.
Steel, Carlos (trans.): "Introduction" to *'Simplicius' On Aristotle On the Soul 2.5–12* (London/Ithaca, NY, 1997).
Stern, Robert (ed.): *Transcendental Arguments and Scepticism: Answering the Question of Justification* (Oxford, 2000).
——, *Transcendental Arguments: Problems and Prospects* (Oxford, 1999).
Stich, Stephen: *The Fragmentation of Reason* (Cambridge, MA, 1990).
Stich, Stephen and Shaun Nichols: "Folk Psychology: Simulation or Tacit Theory?," *Mind & Language* 7 / 1 (1992), 35–71.
Stone, Tony and Martin Davies: "The Mental Simulation Debate: A Progress Report," in *Theories of Theories of Mind*, ed. Peter Carruthers and Peter Smith (Cambridge, 1996), 119–137.

Tappolet, Christine: "Emotions and the Intelligibility of Akratic Action," in *Weakness of Will and Practical Irrationality*, ed. Sarah Stroud and Christine Tappolet (Oxford, 2003), 97–120.

Themistius: *In libros Aristotelis De anima paraphrasis*, ed. Richard Heinze, Commentaria in Aristotelem Graeca 5.3 (Berlin, 1899).

Thom, Paul: *Logic and Ontology in the Syllogistic of Robert Kilwardby* (Leiden, 2007).

——, *Medieval Modal Systems* (Aldershot, 2003).

Thomson, Judith Jarvis: "A Defence of Abortion," *Journal of Philosophy and Public Affairs* 1 (1971), 47–66.

Tittle, Peg: *What if .... Collected Thought Experiments in Philosophy* (New York, 2005).

Todd, Robert B.: "Philosophy and Medicine in John Philoponus' Commentary on Aristotle's *De anima*," *Dumbarton Oaks Papers* 38 (1984), 103–110.

Virgil: *Opera*, ed. Roger Mynors (Oxford, 1969).

Virvidakis, Stelios: "Η Φαινομενολογία της Ηθικής Εμπειρίας στους Εσχατολογικούς Μύθους των Πλατωνικών Διαλόγων," in *Acta. First Panhellenic and International Conference on Ancient Greek Literature (23–26 May 1994)*, ed. J.-Th.A. Papademetriou (Athens, 1997), 525–535.

Walton, Kendall: *Mimesis as Make-Believe. On the Foundations of the Representational Arts* (Cambridge, MA, 1990).

Wardy, Robert: *The Chain of Change: A Study of Aristotle's* Physics *VII* (Cambridge, 1990).

Wilkes, Kathleen: *Real People: Personal Identity without Thought Experiments* (Oxford, 1988).

Williams, Bernard: "A Critique of Utilitarianism," in *Utilitarianism: For and Against*, ed. John Jamieson Carswell Smart and Bernard Williams (Cambridge, 1973), 82–117.

Williamson, Timothy: "Armchair Philosophy, Metaphysical Modality and Counterfactual Thinking," *Proceedings of the Aristotelian Society* 105 / 1 (2005), 1–23.

——, "Replies to Kornblith, Jackson and Moore," *Analysis* 69 / 1 (2009), 125–135.

——, *The Philosophy of Philosophy* (Oxford, 2007).

Wisnovsky, Robert: "Notes on Avicenna's Concept of Thingness (*shay'iyya*)," *Arabic Sciences and Philosophy* 10 (2000), 181–221.

Witt-Hansen, Johannes: "H.C. Ørsted, Immanuel Kant, and the Thought Experiment," *Danish Yearbook of Philosophy* 13 (1976), 48–65.

Wittgenstein, Ludwig: *Wittgenstein's Nachlass. The Bergen Electronic Edition* (Oxford, 2000).

Wodeham, Adam: *Tractatus de indivisibilibus*, ed. and trans. Rega Wood (Dordrecht/Boston/London, 1988).

Zupko, Jack: *John Buridan's Philosophy of Mind*, unpublished PhD dissertation (Ithaca, NY, 1989).

——, "Nominalism Meets Indivisibilism," *Medieval Philosophy and Theology* 3 (1993), 158–185.

INDEX NOMINUM

History of Science and Medicine Library

ISSN 1872-0684

12. Chabás, J. & B.R. Goldstein (Eds.). *The Astronomical Tables of Giovanni Bianchini*. 2009. ISBN 978 90 04 17615 7 (Published as Vol. 10 in the subseries *Medieval and Early Modern Science*)
13. Eagleton, C. *Monks, Manuscripts and Sundials: The Navicula in Medieval England*. 2010. ISBN 978 90 04 17665 2
14. Hill Curth, L. *The Care of Brute Beasts*. A Social and Cultural Study of Veterinary Medicine in Early Modern England. 2010. ISBN 978 90 04 17995 0 (Published as Vol. 11 in the subseries *Medieval and Early Modern Science*)
15. Goddu, A. *Copernicus and the Aristotelian Tradition*. Education, Reading, and Philosophy in Copernicus's Path to Heliocentrism. 2010. ISBN 978 90 04 18107 6 (Published as Vol. 12 in the subseries *Medieval and Early Modern Science*)
16. Holland, J. *Key Texts of Johann Wilhelm Ritter (1776-1810) on the Science and Art of Nature*. Translations and Essays. 2010. ISBN 978 90 04 18367 4 (Published as Vol. 13 in the subseries *Medieval and Early Modern Science*)
17. Streijger, M., P.J.J.M. Bakker & J.M.M.H. Thijssen (Eds.). *John Buridan* Quaestiones super libros *De generatione et corruptione* Aristotelis. A Critical Edition with an Introduction. 2010. ISBN 978 90 04 18504 3 (Published as Vol. 14 in the subseries *Medieval and Early Modern Science*)
18. Kooijmans, L. *Death Defied*. The Anatomy Lessons of Frederik Ruysch. Translated by D. Webb. 2011. ISBN 978 90 04 18784 9
19. Ierodiakonou, K. & S. Roux (Eds.). *Thought Experiments in Methodological and Historical Contexts*. 2011. ISBN 978 90 04 20176 7 (Published as Vol. 15 in the subseries *Medieval and Early Modern Science*)
20. Ariew, R. *Descartes among the Scholastics*. 2011. ISBN 978 90 04 20724 0 (Published as Vol. 1 in the subseries *Scientific and Learned Cultures and Their Institutions*)
21. Shalev, Z. *Geography, Religion, and Scholarship, 1550-1700*. 2011. ISBN 978 90 04 20935 0 (Published as Vol. 2 in the subseries *Scientific and Learned Cultures and Their Institutions*)
22. Roos, A.-M. *Web of Nature: Martin Lister (1639-1712), the First Arachnologist*. 2011. ISBN 978 90 04 20703 5 (Published as Vol. 16 in the subseries *Medieval and Early Modern Science*)

www.ingramcontent.com/pod-product-compliance
Lightning Source LLC
LaVergne TN
LVHW081515300926
849161LV00006B/1185
* 9 7 8 9 0 0 4 2 0 1 7 6 7 *